God, Creation and Climate Change. Spiritual and Ethical Perspectives

LWF Studies, 2009
July 2009

edited by
Karen L. Bloomquist
on behalf of
The Lutheran World Federation

**God, Creation and Climate Change:
Spiritual and Ethical Perspectives**
LWF Studies, 2009

Karen L. Bloomquist, editor
on behalf of The Lutheran World Federation—A Communion of Churches

Copyright 2009 Lutheran University Press and the Lutheran World Federation
All rights reserved. Except for brief quotations in articles or reviews, no part of this book may be reproduced in any manner without prior permission.

Editorial assistance: LWF, Department for Theology and Studies
Design: LWF-OCS
Artwork on cover: LWF-OCS. Photo: Crashing waves at Spoon Bay, New South Wales, Australia. © Brent Pearson

Published by Lutheran University Press under the auspices of:
The Lutheran World Federation
—A Communion of Churches
150, rte de Ferney, P O Box 2100
CH-1211 Geneva 2, Switzerland

This book is also available in Europe using ISBN 978-3-905676-71-6
ISSN 1025-2290

Library of Congress Cataloging-in-Publication Data

Creation and climate change : spiritual and ethical perspectives /
edited by Karen L. Bloomquist ; on behalf of the Lutheran World Federation.
    p. cm. -- (LWF studies ; 2009)
  Includes bibliographical references and index.
  ISBN-13: 978-1-932688-42-9 (alk. paper)
  ISBN-10: 1-932688-42-0 (alk. paper)
  1. Human ecology--Religious aspects--Christianity. 2. Climatic changes--Moral and ethical aspects. 3. Environmental ethics. I. Bloomquist, Karen L., 1948- II. Lutheran World Federation.
  BT695.5.C74 2009
  261.8'8--dc22

                    2009028558

Lutheran University Press, P O Box 390759, Minneapolis, MN 55439
Manufactured in Switzerland by SRO Kundig

Contents

5   **Introduction**
Karen L. Bloomquist

11   **What Do You See, Feel, Believe in the Face of Climate Change?**
An LWF survey (2008)

13   **God, Creation and Climate Change**
A resource for reflection and discussion

27   **Addressing Realities on the Ground**
Colette Bouka Coula

33   **Human Rights and Climate Change**
James B. Martin-Schramm

47   **Who Dies First? Who is Sacrificed First?**
Ethical Aspects of Climate Justice
Christoph Stueckelberger

63   **An LWF Climate Change Encounter in India**

73   **A Faith and Life-Changing Experience**
Anupama Hial

75   **Discerning the Times: A Spirituality of Resistance and Alternatives**
George Zachariah

93   **Caminhada: A Pilgrimage with Land, Water and the Bible**
Elaine Gleci Neuenfeldt

101   **Listen to the Voice of Nature. Indigenous Perspectives**
Tore Johnsen

115   **Wisdom Cosmology and Climate Change**
Norman Habel

127	**Black Saturday**
	Norman Habel
129	**God's Lament for the Earth: Climate Change, Apocalypse and the Urgent Kairos Moment**
	Barbara Rossing
145	**Cross, Resurrection and the Indwelling God**
	Cynthia Moe-Lobeda
159	**Invoking the Spirit amid Dangerous Environmental Change**
	Sigurd Bergmann
175	**Appendix**
	LWF Resolution on Climate Change

Introduction

Karen L. Bloomquist

Today much public attention is being given to "climate change"—the human-caused disruption of the global climate and weather systems, due especially to the emission of greenhouse gases from burning fossil fuels. These gases trap solar radiation entering the atmosphere, thus warming the atmosphere, land and oceans and increasing the frequency and intensity of storms, floods and droughts. Sea levels rise and the coastal lands that are home and provide a livelihood for many millions of people throughout the world, not to mention all the other creatures that depend on such natural habitats, are destroyed.

Climate change is not only an environmental matter but is also correlated with more severe food shortages, loss of livelihoods, conflicts over land and water, increasing impoverishment and the forced migration of peoples, along with other economic and political crises. Some are predicting that without immediate action to redress climate change, fifty years of development gains in poor countries will be permanently lost.

The figure of 250 million who are currently affected each year by climate change related disasters is likely to increase significantly. Climate change is becoming a serious threat to at least half of the world's population. It can no longer be viewed as someone else's problem. The future of life as we have known it on this planet is at stake.

Governments, businesses, communities and individuals need to take timely actions to mitigate or adapt to the effects of climate change, along with churches and other faith communities. The Lutheran World Federation,[1] its related diaconal or development work and a number of member churches are already extensively involved in such efforts.

Yet, underlying all the proposed actions and initiatives is a growing awareness that quick "solutions," as helpful as some of them may be, will not suffice. Instead, major changes are needed in how we think and in what we value. At the 2009 Global Humanitarian Forum meeting in Geneva, where experts discussed their innovative efforts to protect and restore the "global commons" in the face of climate change, it was acknowledged that economic incentives alone are inadequate. They asked,

---

[1] See the appendix for the actions taken at the 2008 meeting of the LWF Council, Arusha, Tanzania.

How can people be persuaded to think beyond their own generation or self-interests? How can the future of the earth itself be part of our thinking? These secular experts were posing the very kinds of questions that are deeply spiritual and ethical in nature. In other words, climate change has opened up the theological-ethical agenda that churches and other faith communities urgently need to take up.

Awareness of climate change is provoking old and not so old questions about the relationships (a) between human beings and the rest of creation; (b) between God and nature; (c) between divine activity and human responsibility; and (d) among human communities globally. Much of what in the past were referred to as "acts of God" are now seen as caused at least in part by human activity. Climate change may literally be melting icebergs but it also exposes metaphorical "icebergs" of how God, human beings and the rest of creation have been conceptualized in ways that contribute to the injustices that are escalating under climate change.

People in local communities are likely to draw upon a variety of spiritual resources—including local or indigenous wisdom and practices—for coping with or adapting to what is occurring. For Christians, biblical resources are likely to be prominent among these. Attention needs to be given to how we read and interpret the Bible in relation to what we are experiencing—not with a sense that God is punishing or abandoning us, but with a sense of God's abiding promise which empowers us to act. We need to go beyond the poles of either a fearful sense of apocalyptic doom that only waits for God's inevitable judgment or, on the other hand, a moralistic sense of activism, driven by a sense of what we need to do to "save" the world. God has already saved the world. The question is, How do we participate in the redemption of all creation to which Scripture testifies, and embody hope for the future rather than succumbing to despair?

In this book

Recent LWF work on climate change, which the LWF Council called for in 2007, began with the development of a survey that asked people in local communities what they see, feel and believe in the face of what is occurring. Select responses to these basic questions informed the development of a resource for reflection and discussion in local settings, which is reprinted here and available in English and German at **www.lutheranworld.org/ What_We_Do/Dts/Programs/DTS-Church-Social_Issues.html** (with discussion questions and photos). The premise behind this resource is that climate change is provoking the need to reconsider and revise much of what people have previously assumed or believed.

In October 2008, the LWF Department for Theology and Studies convened a small consultation of biblical scholars, theologians and ethicists who have been active in this area. They were asked to develop more in-depth papers related to what was presented in summary fashion in the above resource. Most of the articles in this book are by participants in this consultation.

Colette Bouka Coula, Cameroon, staff member of the LWF Department for World Service in Geneva, cites some of the ways in which DWS related programs have been working to mitigate and adapt to climate change in some of the most vulnerable areas of the world, since long before these efforts were associated with "climate change."

James Martin-Schramm, who teaches at Luther College in Decorah, Iowa (USA), summarizes the key information from the Intergovernmental Panel on Climate Change and connects climate change to the human rights concerns that have been on the ecumenical and civil society agenda for some time. The Swiss Protestant ethicist, Christoph Stueckelberger, director of Globethics.net, systematically takes up different ethical aspects of climate justice, and the difficult tensions at stake. Furthermore, he reflects on his learnings from over twenty years of involvement in these matters through the World Council of Churches and other organizations, and urges collaborative work with many other actors in society.

The commitment to focus on where people, land and water are especially vulnerable to climate change was carried forth in a special LWF encounter in April 2009 in Puri (Orissa), India. Here participants witnessed the dramatic effects of how climate change has already washed away land and villages, and how the people themselves are responding. The communiqué, "An LWF Climate Change Encounter in India," poignantly describes what the participants witnessed, and their call to the rest of the Lutheran communion. One of the participants, Anupama Hial, a pastor in Orissa, shares how she was impacted by the encounter. George Zachariah, who teaches at Gurukul Lutheran Theological College in Chennai, India, deepens and broadens the ethical analysis of climate change and its interconnection with other realities of injustice, and calls for resistance to what is occurring and the development of alternatives.

The Brazilian biblical theologian, Elaine Gleci Neuenfeldt, who currently staffs the women's desk in the LWF Department for Mission and Development, proposes a pilgrimage theology "on the road" which, in the face of climate change, focuses on experiences of the Brazilian women's movement in relation to the land and water, and how this resonates with and is empowered by biblical accounts of land and water. The Sami pastor, Tore Johnsen, from Norway, raises up the Sami *yoik* that protests against how the land was used by the intruding culture, and develops the "circle of life" indigenous spiritual tradition shared also with Native Americans. On this basis, he reconceives central aspects of Christian theology in ways that are more connected with the earth and cosmos.

Norman Habel, an Australian Lutheran Old Testament scholar, who for many years has promoted reading the Bible from the perspective of the earth, draws upon the biblical wisdom tradition, with its underlying design and innate codes for all of life, which in human folly we have broken and which results in climate change. He also connects this with the ravaging bushfires in Australia, another result of climate change. Barbara Rossing, who teaches New Testament at the Lutheran School of Theology in Chicago (USA), and regularly speaks publicly to counter the apocalypticism some associate with climate change, asks where God is in this crisis. In interpreting Revelation and other biblical texts, she finds that God is lamenting with us and the earth, rather than punishing or cursing us. "God is crying for a world that needs to be freed from the toxic system of imperial exploitation." She calls for public *metanoia* at this urgent kairotic time. Both Habel and Rossing were Bible study writers for the 2003 LWF Assembly and participants in the 2009 Puri encounter.

The moral inertia that relatively privileged Christian feel in the face of climate change is the point of departure for Cynthia Moe-Lobeda, a Lutheran ethicist who teaches at Seattle University (USA), and who has also written for other LWF resources. She draws upon Luther's (and Bonhoeffer's) interpretations of a theology of the cross, as well as Luther's notion of the indwelling God, to empower moral agency so that we might turn from being creation's "uncreators" to working for more just, sustainable ways of life for all.

In the final article, Sigurd Bergmann, who teaches at the Norwegian University of Science and Technology in Trondheim, claims that environmental changes are radically changing culture, religion and the conditions for faith, and explores where the Spirit of life is "taking place" today. Amid a sweep through history, intellectual thought, art and current technological and economic realities, he points to the distinctive task of Christian theology in relation to these other perspectives, especially in remembering the suffering and liberation of creation. He concludes by suggesting that the church be reimagined "as an agglomeration of local places and a global space for creative experiments in the arts of survival in environmentally changing contexts."

Where from here?

It is hoped that the articles in this book will stir and motivate you not only to face the urgency of the unprecedented global crisis of climate change as manifest in your context, along with the various social, economic and political issues related to it, but to reflect in new ways on the spiritual resources that are a part of Christian and other traditions and are so crucial for empowering and sustaining us in facing and redressing this challenge not only now but also for the future.

We are eager to hear your further thoughts and examples of how you are affected and what you are doing: Contact **kbl@lutheranworld.org**

What Do You See, Feel, Believe in the Face of Climate Change?

An LWF survey (2008)

What is different today? In recent years, what general changes have you noticed in the climate in your area? How is this affecting the land, the plants, the air, the animals and the people? What is different from what your parents or grandparents experienced?

Who? Who or what is especially affected by these changes? Who especially bears the burden? Who or what is especially responsible for climate change?

Why? How do people explain these changes? Why are they happening? (The stories or folk wisdom as well as more scientific explanations.)

What has gone wrong? In the relationship between human beings and the rest of creation? In the relationship between people? In the relationship with God?

God? How do you feel God is related to or involved in this? What questions would you pose to God? How is your faith in God affected? What spiritual resources do you draw upon?

The future? How do you view the future, for your community, coming generations, and the earth as a whole? What do you fear or hope for? What spiritual resources do you draw upon?

Solutions? What needs to change in your society? What trade-offs are there? What is being done that can make a difference? What local solutions would you propose?

God, Creation and Climate Change

A resource for reflection and discussion

What is going on?

Around the world we are experiencing the effects of climate change: water and air temperatures are rising at alarming rates, adversely affecting the habitats that sustain life for fish, animals, plants and human beings. Devastation caused by more severe droughts and floods is increasing. Storms and hurricanes are becoming more frequent and intense. New diseases are appearing and old ones are spreading. For example, because of warmer temperatures the breeding of malaria-carrying mosquitoes has increased. In overly industrialized areas, the air quality is deteriorating. Climate conditions are affecting people's health and in some areas heat-related deaths are on the increase.[1] Hunger is predicted to escalate as the climate changes.

The predictable, dependable order of things is changing: when winter or summer begins, or when the rainy season comes, if at all, is becoming ever more unpredictable. The availability of clean water to sustain life is jeopardized, especially as much of it is being privatized. Houses built on what seemed to be solid ground are suddenly swept into raging waters. Growing seasons for crops are changing significantly, as is the yield of crops related to soil quality, moisture and erosion. In some places, winters are becoming colder, and in others, warmer. Where the food needed for daily life will come from, and when, is becoming more unpredictable, making the right to food more precarious, especially for the most vulnerable.

Some are wondering whether they can still rely on God's promise to Noah: "As long as the earth endures, seedtime and harvest, cold and heat, summer and winter, day and night, shall not cease" (Gen 8:22).

As the Intergovernmental Panel on Climate Change (IPCC) concluded in 2007:

> Human beings are exposed to climate change through changing weather patterns (for example, more intense and frequent extreme events) and indi-

rectly through changes in water, air, food quality and quantity, ecosystems, agriculture, and economy... Increased frequency of heat stress, droughts and floods negatively affect crop yields and livestock beyond the impacts of mean climate change, creating the possibility for surprises, with impacts that are larger, and occurring earlier, than predicted using changes in mean variables alone. This is especially the case for subsistence sectors at low latitudes. Climate variability and change also modify the risks of fires, pest and pathogen outbreak, negatively affecting food, fiber and forestry.

In other words, the predictabilities on which we have depended for life as human beings have long known it are changing dramatically. We wonder on what we can depend for the future.

As numerous studies have indicated, it is especially human activity that is causing or at least significantly contributing to climate change. Nevertheless, for people in many parts of the world for whom there is a close relationship between the divine and what occurs through nature, the "God questions" cannot be ignored.

God and climate change?

Some people view climate change as if God had disappeared from the scene, had been pushed to the margins by human activity and was no longer active in the cosmos. But for persons of faith, the extensive global and cosmic realities of climate change need to be considered in light of how we understand God, creation and humanity.

In many passages of the Bible, natural occurrences such as those occurring today due to human-induced climate change, were attributed to God. People in many parts of the world still do so today. God has been considered the agent causing floods, storms, droughts and other local and global "natural" catastrophes. People view what is occurring as being acts of God, and ask why.

Throughout the ages and from different faith perspectives, weather related disasters have often been considered as "acts of God." When the destructive effects of climate change occur, some immediately respond that God must be punishing human beings—and this is how they interpret certain biblical passages. People are told simply to wait and endure God's judgment, rather than doing anything to change what is considered to be God ordained and thus, inevitable.

As people of faith, we maintain that somehow God is involved in climate change—especially to wake us up to the urgency of what is occurring—but we cannot attribute climate change only to "acts of God." We must also turn to science, through which we learn more deeply, and with greater awe, about what God has created.

Many of the problems associated with climate change have arisen because of how human beings have misused that which God has created for the benefit of all creatures. The church has long taught that we are to be good stewards or caretakers of what God has given, and must continue to do so. But the challenge goes deeper than this.

To a large extent, many global facets of the climate change crisis have come about because of how interrelated assumptions about God, creation and human beings have profoundly influenced and shaped modern societies, institutions and ways of life. These have been passed on through centuries of teachings in the church, which for example, separated nature from grace. Western thinking which tended to separate human beings from the rest of creation contributed to the rise of industrialization and capitalism. Developments such as these, in turn, have spread throughout the world. Over the centuries, these assumptions, and the practices based on them, have contributed cumulatively and now disastrously to climate change, which seriously threatens the future of life on the planet as we have known it. The effect of climate change is like that of hunger—it weakens, gnaws and although it may not be the sole cause of death, it pushes you in that direction.

These assumptions include,

- That God is transcendent, unchanging, all powerful, a heavenly monarch or patriarch ruling above and controlling the world, untouched by earthly realities

- A worldview with God at the top, then men over women, children, animals and, at the bottom, the rest of creation

- That as agents of God, human beings are to use or exert power over the rest of creation

- That God acts primarily in history and not also in and through nature

- That only human beings, and specifically Christians, benefit from God's grace or redemption

- That spiritual matters are separate from what is embodied or "of this earth."

The influence and effects of assumptions such as these have spread over the entire world through colonization, conquest, empire building, missionary movements and economic development. This continues today through accelerated processes of globalization. These assumptions have undergirded and furthered habits and practices around the world that we now realize have, over time, contributed dramatically to climate change and are threatening life as we have known it.

Such practices include,

- Economic life based on endless quests for ever greater growth and profit driven by greed, which the global economic crisis is starkly exposing today

- Increasing dependence on fossil fuel extraction to further this development

- Conquering practices of colonization and empire, especially in the constant quest for more resources and markets

- Patriarchal ideologies that perpetuate control over and oppression of both women and the earth

- Discrimination against all those seen as "other" because of their gender, race, ethnicity, caste, economic or political status

- Assuming that some aspects of creation (such as trees, water or air) are dispensable, rather than respecting and valuing all of creation

- An anthropocentrism that tends to value only that which serves human ends.

Climate change is provoking the need for climatic changes in some faith understandings that have long been taken for granted. Climate change may literally be melting icebergs but it also exposes metaphorical "icebergs" of how God, human beings and the rest of creation have been conceptualized in ways that contribute to the destruction and injustices that have escalated under the currently reigning regime of climate change.

The Triune God is intimately related with all of creation

When people think about "God" they often refer to a supreme being who reigns over and above the world as an almighty ruler or monarch (almost always as "he"). When something goes wrong in nature, such as occurs under climate change, it is then immediately assumed that this is caused by "God"—as an almighty actor standing outside of and controlling all that occurs on earth. Throughout the ages, and in many religious traditions, humans have prayed and offered sacrifices so that God would bestow favorable conditions for growing crops, protect from storms and rising waters, and control the natural forces of the environment. After all, isn't God the power over all the cosmos, and thus the One able to control everything, including climate change?

Many biblical references seem to reflect such understandings of God. These are often interpreted in ways that make too sharp a separation between God and nature. In part, this was to distinguish ancient Israel's understanding of God from some of the nature religions, according to which the fate of humans was determined by the gods acting in the cycles and forces of nature. But, making a sharp separation between God and nature becomes a problem when it overlooks the intimate relationship that God has with all of creation, as described in the beginning of Genesis and in many other places in the Bible.

"Holy, holy, holy is the Lord of hosts, the whole Earth is full of his glory" (Isa 6:3) The glory here is the vibrant presence of God, which was earlier depicted as the fire cloud of God's presence at Sinai. Later it "filled" the tabernacle and then the temple of Solomon. But here Isaiah goes further and declares that God's very presence fills the whole earth, which is God's sanctuary.

The God revealed in the Hebrew Scriptures is not unchanging in the same way as are some other gods. God is related to creation and history not by being immune to space and time but by keeping promises. "God's will" should not simply be equated with natural occurrences, insisting that God is causing all that occurs. Yet, at the same time, we may glimpse what God has created and intends, which contrasts with the breakdown or destruction of the fragility of creation that is occurring through climate change. Creation is good because God created all that is, although not everything that occurs in creation is good.

In the Book of Job, when Job reaches the depths of despair he not only accuses God of harassing humans unjustly, but also indicts God for God's rough treatment of creation. Job claims that God uses divine wisdom to hold back the waters until they dry up and to unloosen them so they flood the land (Job 12:15). In chapters 38–39, God takes Job on a tour of the various aspects of the cosmos to enlighten him about the mysterious "ways" of the natural world. It is not for Job to try to rule nature, but to explore how God has created all that exists and to discover how humans fit into this mysteriously complex design of God.

Here and elsewhere in Scripture, we begin to catch a new sense of who God is—not an all-controlling monarch who punishes even the innocent, but God revealed yet hidden throughout creation. God's grace and love are ultimately more crucial than might and power. God is intimately related with humans and the rest of creation, present in the midst of vulnerability and suffering.

Today, a similar shift is called for in how we imagine or think of God, standing as we do in the midst of a creation suffering the effects of climate change. Those who have used little of the earth's resources find themselves the most dramatically affected. Yet, blaming God for this is not the answer. As Scripture continually reminds us, human unfaithfulness to God is the problem. This is reflected in the unjust treatment of humans and the rest of creation.

The twentieth-century Lutheran ecological theologian, Joseph Sittler, insisted that nature comes from God, cannot be apart from God, and is capable of bearing the glory of God.[2] Grace is the fundamental reality of God, as Creator, Redeemer, Sustainer. Grace is the sheer giveness of life, the world and ourselves. We are "justified" by grace even in our relation to the things of nature. Condemnation (the opposite of justification) is present in the absence of a gracious regard for nature, such as when we pollute or use nature as a dump.

This concurs with Martin Luther's sixteenth-century perspective: all of creation is the abode of God. Rather than removed or set over creation, God is in, with and under all that is creaturely. Despite all the negativities—such as the disruption and destruction occurring due to climate change—we still trust that God is at work in this world, often hidden beneath its opposite.

This is also at the heart of what Luther meant by a theology of the cross: God is neither to be seen nor sought behind creation nor inferred from it, but only recognized in and through it. The cross reveals how radically God is immanent in creation.

Throughout the history of the church, there have been many debates as to what is most central about God. For some, God's almighty power has been key, while for others it is God's everlasting love. For Lutherans and many other Christians, what is most important is that God is love. God seeks to be in intimate relation with what God has created, including human beings: being **with** rather than being **above** or distant from creation.

It is the Spirit of God (*ruach*) who conveys this sense of intimacy between God and creation. God is alive and active as the Spirit, giving life to all that is. God's "breath" expresses God's life-creating, life-preserving goodness.

The Spirit of God is the inexhaustible, ever-creative power of God, the life-bestowing force of creation and re-creation, ruling not by controlling power but through powerlessness. God overturns our human notions of power. God's transforming activity goes beyond any human-erected boundaries, and cannot be limited by dominant values and systems, such as those that have contributed to climate change. The Spirit of God helps human beings to perceive God in the midst of creation.[3] God rules through seeming powerlessness.

In its confession of faith in the Triune God, the church has insisted that God is essentially relational, not an autonomous God but God-in-communion. This is in sharp distinction to views that consider God to be a being who is self-sufficient and separate from creation, controlling it from "outside" or "above," as does an imperial ruler. God who is love seeks to be close to, not distant from creation.

The purpose of Trinitarian theology is not to define God or God's "substance," but to describe the whole, interrelated gracious movement of God who seeks communion—intimate relationship—with what God has created. Creation is far more than just a backdrop for God's main redemptive activity in human history. It is the redemption of **all** creation that is at stake (Rom 8), not redemption *from* creation.

> God's labors of creation, preservation, and redemption are not three separate or separable works but a single labor, whose object is precisely the birthing of the world that God intends. God is "in labor" in the world, for the world, that it might become what, in its conception, it is.[4]

In other words, God is the source, power and goal—the spirit that enlivens the complex processes of creation. God is the source of all being rather than one who intervenes from outside. This is how theologians such as Sallie McFague refer to God: as the inspirited body of the whole universe, creating, guiding and saving all that is. Rather than assuming God to be like a will or intellect ordering and controlling the world, God is the breath that enlivens and energizes the living breathing planet. God permeates, suffers with and energizes the innermost aspect of all that is created, in ways known and unknowable, in ways that are both intimate and transcendent.[5] We can only gratefully receive rather than solve this mystery.

Picturing God's activity in such organic ways is more appropriate than in machine like ways, which have compounded the problems we face today. The machine model assumes that rational control is what is important, with God as the ultimate fixer. Instead, the focus shifts from control to relationships—interdependent relationships throughout all of creation.

This is similar to how many indigenous traditions and faiths have viewed the relationship between God and creation. The interdependence of everything has been common knowledge throughout most of world history—all the relationships necessary for life to flourish, including the predictability of the climate. Many indigenous peoples have long assumed such an ecological vision of life, in contrast to perspectives which value human life at the expense of other forms of life.

Taking creation seriously as God's abode means that the physical space of creation becomes important. This spatial dimension has long been celebrated, for example, in the Psalms: "How lovely is your dwelling place, O Lord of hosts! … even the sparrow finds a home … at your altars." (Ps 84:1–3). We dwell in God who surrounds us, from before and beyond all time: "Lord, you have been our dwelling place in all generations. Before the mountains were brought forth, and ever you have formed the earth and the world, from everlasting to everlasting you are God" (Ps 90:1–2).

The incarnation—God becoming fully human in Jesus of Nazareth—is the clearest testimony to God's intimate relationship with what is created. In him, divinity and humanity, heaven and earth are brought together. The central festivals of the church year emphasize this in powerfully poetic and symbolic ways. At Christmas, "heaven and nature sing" as a bright star in the heavens is linked on earth with a lowly manger. On Good Friday, God is revealed in the One who suffers and dies with all of creation, and at Easter, heaven and earth exult with the living God. At Pentecost, the wind of the Spirit blew from heaven, empowering those in the early church to communicate across their earth-bound differences.

So what about human beings?

In recent centuries in the West, and throughout much of the world today, the above perspectives have been overshadowed. Some human beings have acted as if they were demigods who can order and control, for their own self-interests, the land, trees, air, water and other creatures, including vulnerable human communities. This often occurs in the name of "development" or "progress." The air, water, soil and plants are valued in so far as they will further human development or progress, rather than because of their own intrinsic worth. The accumulation of money and goods has displaced the liberating economy of the Creator, based on synergy, cooperation and life-enhancing justice for all of creation.

Consequently, the delicate interrelationships within creation have been upset. Creation's protest is now being experienced through climate change.

Being creatures within creation is at the core of a Christian anthropology. However, many human beings have lost the sense of being part of a living, changing, dynamic cosmos, which has its being in and through God.

Based on the two creation stories in Genesis 1 and 2, human beings have often assumed themselves to be the crown of creation, or the main purpose for which God created everything else. This has been due to misunderstanding the call to "have dominion over" (Gen 1:28) in ways that have led to the exploitation of creation, rather than a sense of responsibility and accountability for what God has created. In Genesis 2, in the midst of the plants and water of the garden, God forms the first human being from the dusty earth and breathes life into *'adam*. Tilling and keeping the garden—cultivating and preserving God's creation—is the mandate given to humans. Human beings are to be servants of the rest of creation, not its rulers. This is similar to how in Mark 10:41–45, Jesus calls the disciples to follow him by serving rather than ruling over others.

Assuming human beings are separate from or above nature can imply a complete freedom of action toward creation—using or exploiting it in ways that serve human ends, or as "raw material for human sustenance and aggrandizement."[6] Instead, creation has a dignity and purpose that goes beyond human purposes.

Sin and salvation are both spiritual and earthly matters; they have to do with how we relate to the forms of God's presence we encounter in our daily, ordinary lives.[7] Sin is our failure to live out of the relational matrix we share with the rest of creation and with God. It is our refusal to remove ourselves from the center of the world. We attempt to escape our creaturehood and the relationships and vocation that belong to it. Sin is living falsely, contrary to the appropriate relationships that constitute reality. When relationships are violated, injustice, abuse and destruction result. Sin is refusing to accept the limits and responsibilities of our place within the whole of creation.

Environmental exclusion in the form of exile is a core theme of the Old Testament, and it speaks to the condition of those millions who are already finding they are forced to migrate from their ancestral lands because of drought and flood caused by climate change.[8]

The writings of the Old Testament Prophets repeatedly remind us that God will not tolerate injustices inflicted on other human beings and on the rest of creation, through dominating power, control and oppression. However, in many of these passages where God responds to injustice, God is depicted as an all-controlling male ruler or warrior who acts in punitive, violent, destructive ways. The problem is that this legitimizes rather than transforming patterns of violence against humanity and creation.

The power to change the injustices should be consistent with God's overall purpose of restoring and transforming creation. Carol Dempsey indicates how this is conveyed in especially chapters 42, 49, 52, 53, 61 and 65 of the Book of Isaiah:

> (1) The redemption of humankind is connected to the restoration of creation; (2) the human community has a responsibility toward all creation; (3) the vision of Isaiah 65:17–25 can no longer remain apocalyptic or eschatological but must become a reality for the planet and life on the planet; (4) the divine vision for all creation is one that speaks of respect for all of life and life lived in balance and relationship….The focus must shift from the use of power to dominate, control and oppress to the use of power to empower oneself and others and liberate all of creation from its groaning and oppression.[9]

The call to repentance in Mark 1:15 can be heard as a call to return to a proper relationship with the Creator and creation, "a call to be liberated from our human perceived need to be God, and instead to assume our rightful place in the world as humble two-leggeds in the circle of creation with all the other created."[10]

Given the kairos of climate change today, there is an urgent need for repentance or conversion.

We need to shift from:

- Human independence, *to* human interdependence with the rest of creation

- Making separations based on oppositions and dualisms, *to* emphasizing interrelated balances and connections

- Technological control, *to* respect for, care and balanced use of creation and its resources, including through appropriate technologies

- Creation as only the backdrop for human worship, *to* creation pulsating with life, pathos and worship of God

- An exclusive focus on God active in human history, *to* God active in, with and through the spatial realities of the whole creation, in which humans participate

- A predominantly Christocentric focus on the redemption of human beings, *to* Trinitarian thinking that takes more seriously creation, the Spirit and the interrelationships throughout the cosmos, with all of creation as the scope of redemption

- Sin only as a broken relationship between humans and God, *to* the sinful ways relationships with creation are broken

- God's grace separate from nature, *to* God's grace known in, with, through and transformative of nature

- Transcendence that is spiritualized and removed from the life and matter of creation, *to* a sense of the divine mysteriously active in, with and through what is created

- An obsession with progress and development as measured in economic terms, *to* what will result in more sustainable life for all of creation

- Allegiance to the global market system, *to* being inspired by a vision of God's economy for the sake of the well-being of all, including earth itself

- A focus only on technological or market-base "fixes," *to* the healing of creation.

The redemption of all creation

> God's anger leads not to judgment but to redemption, not just of human beings, but of all creation: "the creation itself will be set free from its bondage to decay and will obtain the freedom of the glory of the children of God" (Rom 8:21). Because of God's transforming grace, rather than because of fear, we are empowered to change our attitudes, lifestyles, and practices—to put things right again. The way things are now cannot continue with "business as usual." Instead, the God of grace who is active through, with and in nature, is revealing how urgent it is to recover the spiritual significance of valuing our common good with the rest of creation.[11]

In the fourth century, St Ambrose wrote, "For the mystery of the Incarnation of God is the salvation of the whole of creation."[12] Salvation is the direction of creation, and creation is the place of salvation.[13] In other words, the health and well-being of all of creation is what salvation is about. Christ's liberating, healing and inclusive ministry takes place in and for creation. In Christ, God identifies with all suffering bodies, including the suffering of creation itself.

This cosmic scope of Christ is communicated especially in Colossians 1. The horizon of salvation or redemption or reconciliation is widened significantly here. Its focus is not on human beings; in fact, they are not even mentioned in this passage. Instead, celebrated here is the intimate relation of Christ and the whole of creation, from before the dawn of time. The fullness of God comes to dwell bodily in creation. The powers of this world are put in their place, and broken relationships throughout creation are restored or reconciled.

Similarly in Romans 8, salvation not only includes human beings but the whole cosmos. Creation itself longs for the revealing of those who, through the power of the Spirit, will rescue the whole created order, and bring about that justice and peace for which the whole creation yearns. This builds on the biblical promise of a new heaven and earth (Isa 65 and 66) and on the creation story in which human beings are to be caretakers of creation. The freedom for which creation longs will come about through human agents, transformed by the Spirit, to bring wise, healing and restorative justice to the whole creation.[14]

In the earthly life of Jesus as recorded in the Gospels, we see one who was continually challenging the traditional dualisms by which people lived their lives: male over female, rich over poor, humans over nature. His compassionate love and justice embraced all of creation, leading him to cross all kinds of boundaries of his day.

Similarly, climate change transgresses boundaries, of both natural and human-defined separations, of communities, of nation-states, of lands, of waters, of near and distant neighbors, of rich and poor, of different cultures, of the past and the future. Many of its effects know no boundaries. Climate change reminds us that we are all in this together. It is the future of life on the planet that is at stake. Yet some bear the brunt and the consequences far more than others, and are far more vulnerable. Under climate change, nature has become "the new poor," as vulnerable and expendable as poor human beings and communities have been. This is where our attention and priority especially needs to be.

The church is far more than a just another actor in civil society for addressing climate change. It has a global even cosmic expanse, crossing boundaries of both space and time. It includes those who are contributing most dramatically to climate change as well as those rendered most vulnerable by it; together they are interconnected and transformed into each other, as members of one communion. The communion of saints crosses all boundaries of time—those in the past and in the present as well as future generations whose very possibilities for life are being jeopardized by climate change.

Furthermore, through the Sacraments, God's promises become tangible through common elements of creation—water, bread, and wine—through which we are redeemed, nourished and empowered. We are redeemed by God not apart from but through what is created. We have been washed in the waters of redemption in baptism and fed with the bread and wine of Holy Communion. Through these Sacraments, the life-sustaining power of God's promises is effected in us, as a foretaste of the feast to come. The church bears witness to the new creation, as a communion, as the body of Christ in the world that God has created and will bring to fulfillment.

Living out of this present and future reality, Christians should be at the forefront of redressing the effects and changing the course of climate change.

- We are challenged to see new possibilities for reconciliation and restoration within creation, in ways that will benefit all rather than just a few.

- The reality of God's redemption is lived out as we pursue greater justice for all. It does not suffice to address the crises evoked by climate change through short-term fixes or "solutions" that only reflect the same old paths of economic and human progress which have brought us to this point.

- We must move beyond narrow anthropocentric views of life, and embrace more interconnected views in which God, human beings and the rest of creation are intimately related.

- When we do so, the injustices imposed on other communities or other realms of creation become all too apparent, as well as our capacity for putting things right again, in communion with the rest of creation.

Notes

[1] At **www.windows.ucar.edu/tour/link=/earth/climate/cli_effects.html&edu=high**.

[2] Steven Bowman-Prediger and Peter Bakken (eds), *Evocations of Grace: The Writings of Joseph Sittler on Ecology, Theology and Ethics* (Grand Rapids: Eerdmans, 2009), p. 104.

[3] Michael Welker, *God the Spirit* (Minneapolis: Fortress, 2004), p. 334.

[4] Douglas John Hall, *Professing the Faith* (Minneapolis: Fortress, 1993), p. 167.

[5] Sallie McFague, *The Body of God: An Ecological Theology* (Minneapolis: Fortress, 1993), esp. pp. 145, 147.

[6] Hall, *op. cit.* (note 4), p. 167.

[7] McFague, *op. cit.* (note 5), p. 114.

[8] Michael S. Northcott, *A Moral Climate: The Ethics of Global Warming* (Maryknoll: Orbis, 2007), p. 161.

[9] Carol J. Dempsey, *The Prophets: A Liberation-Critical Reading* (Minneapolis: Fortress, 2000), p. 179.

[10] George Tinker, *Spirit and Resistance: Political Theology and American Indian Liberation* (Minneapolis: Fortress, 2004), p. 113.

[11] Rolita Machila, "Why are Earth and God Angry?," in *Thinking it Over . . .*, Issue 20 (August 2008), at **www.lutheranworld.org/What_We_Do/Dts/DTS-Welcome.html**.

[12] Exposition of the Christian Faith, V, VIII, 105b.

[13] McFague, *op. cit.* (note 5), p. 180.

[14] N. Thomas Wright, in *The New Interpreter's Bible*, vol. 10 (Nashville: Abingdon 2002), pp. 596–97.

Addressing Realities on the Ground

Colette Bouka Coula

As the relief and development arm of the Lutheran World Federation (LWF), the country, regional and associate programs of the Department for World Service (DWS) have long been working in contexts where people and the rest of creation are especially vulnerable. Responding to and preparing for disasters and creating sustainable communities were priorities long before the relation to climate change became evident.

Responding to climate change is part of reacting to the struggles of those who are trying to survive in very difficult settings. Battles are fought over the right to food and access to basic social services, to be accepted and to live as persons infected or affected by HIV and AIDS, as well as how to survive both natural and human-caused calamities and to protect and take care of the surrounding environment.

Climate change and its effects deeply affect people's livelihood. Therefore, alternative ways have to be identified together with the local communities in order to cope with and adapt to the situation. Actions carried out through a number of DWS projects are designed to mitigate the effects of climate change affecting communities in intervention areas. This has been occurring for many years, although often under terms other than "climate change."

Moreover, DWS engages in advocacy. For instance, in Central America, social movements opposed to the policies of governments and transnational corporations are supported. In Zimbabwe, a new climate change project is being initiated that aims at contributing not only to local level environmental management, but also to the development of a national policy on climate change.

In the following, I shall focus on some aspects of what LWF/DWS country programs have done and are doing together with the local community. Two aspects of interventions under emergency response are highlighted here: disaster risk management and environment protection and conservation, which are directly linked to climate change. The information listed here is based on reports by the respective country programs.

Emergency response and disaster risk management

Beside the emergency operations carried out with refugees, Internally Displaced Persons (IDPs) and victims of wars and other social disturbances, climate change induced by natural disasters, including outbreaks of epidemics, are new challenges for some communities and ongoing ones for others (such as in Angola, Bangladesh, Cambodia, etc). Climate change manifests itself in disrupted seasonal weather patterns causing intermittent and cyclical floods, droughts and cyclones. Drought and crop failure result in an increased number of persons seeking food relief (Swaziland). Floods, leading to malaria, are becoming more widespread (Zambia).

Responses are varied but are mainly based on building and strengthening the capacities of the local communities to deal with the situation. This includes,

- Angola: training communities in cholera prevention and staff in disaster preparedness

- Bangladesh, Balkans, Cambodia, Central America: training communities in disaster preparedness

- Cambodia: establishing community-based disaster risk management training to help communities prevent, prepare for and mitigate natural disasters

- Central America: training local risk management committees and establishing early warning systems

- Mozambique: strengthening local structures such as village committees for disaster preparedness through training

- Nepal: mobilizing disaster management teams at the community level, providing training in small-scale flood intervention measures and increasing the community's capacity effectively to respond to disasters

- Tanzania: training or capacity building for disaster response, creating awareness to enhance communities' capacity to respond and adapt to climate change

- Zimbabwe: facilitating development of disaster preparedness plans for communities.

According to the different country reports, the impact of the disaster risk reduction programs is felt in the communities after these training programs have been conducted. Other ways of responding are through relief food distribution, training communities in agricultural practices, introducing more drought resistant and flood tolerant crops and distributing mosquito nets in situations of malaria outbreaks.

As a result of climate change, natural resources such as firewood are becoming scarce which, in turn, can lead to conflict. In Ethiopia, the provision of water that can be managed by communities, soil and water conservation strategies, conflict resolution and peace-building initiatives have been helpful in furthering community cohesion.

The necessary humanitarian interventions have taken the priorities and choices of the communities into consideration. Even under such conditions, communities have been sensitized to human rights issues, the right to relief and to development. This has helped communities to break out of the vicious circle of poverty and vulnerability. Preparedness and mitigation systems are put in place with the communities that participate in analyzing vulnerability and devising mechanisms to combat disasters.

At the community level, the contribution of women to disaster management is vital. Training has brought about significant changes in the attitude of communities prone to and affected by disasters: they now prepare themselves before disaster strikes again. In India, acquired skills helped disaster management teams to help their own communities and neighbors that were less well equipped and skilled.

Environmental protection and conservation

Global warming is accompanied by deforestation, soil erosion, land and soil degradation, loss of soil fertility and biodiversity, increasing soil salinity, drought and a scarcity of wood. In Central America, water resources are depleted through irrational use by governments and transnational corporations, the development of mining enterprises and non-sustainable logging. There is uncontrolled logging, destruction of endangered plant species, and excessive land use leading to a shortage of firewood (Mozambique).

With the support of DWS country programs, communities have responded to these challenges in various ways.

- Angola: communities have been able to identify alternative technologies such as making bricks out of anthills and building with bricks rather than with wood.

- Bangladesh: mitigating actions, including training on environmental issues to build communities' awareness, environment education and rehabilitation, have been carried out. As a result, communities have been engaged in establishing tree nurseries and tree planting and reforestation projects to fight against erosion that threatens farmland. For instance, millions of trees have been planted.

- Bangladesh: in limited ways, RDRS Bangladesh is extending support to managing garbage collection of the municipal area of the Rangpur district. Garbage is transformed into manure, and communities are involved in compost preparation to replace chemical fertilizer. Similar projects are carried out in India, Liberia and Rwanda.

- Bangladesh and Rwanda: biogas plants and solar lightning are introduced in rural areas and solar lamps used in refugee camps in Nepal.

- Burundi, Eritrea, Zambia and Zimbabwe: development of fuel-saving stoves and the increased use of local smokeless stoves by the communities to reduce the consumption of firewood. In Rwanda, fuel-efficient stoves, promoted by the program, have been adopted by the government and are now used nationwide.

- Cambodia: organization of community forestry projects and promotion of alternatives to charcoal and firewood selling have been adopted as a means of livelihood.

- Ethiopia and India: well and spring construction, water harvesting (or collecting) structures are built.

- Haiti and Peru: protection of water sources and the few remaining forests, as well as replanting native pasture grasses.

- India and Rwanda: farmers are reverting to traditional crops that require less water and are more pest resistant, and efforts are undertaken to develop seedlings better adapted to local conditions.

- India: the environment is protected, conserved and regenerated, the community is sensitized, Indigenous knowledge is drawn upon, and environmental friendly practices are carried out; disaster management teams are organized and trained.

- Mauritania: soil and water conservation activities are taking place, as well as rehabilitation of water retention structures by affected communities, resulting in increased in soil fertility, expansion of cultivable areas, recharging of water ground.

- Mauritania: promotion of biological agriculture, use of biological treatment methods in gardening.

- Nepal: improved seasonal and off-season farming techniques are used. Bioengineering, rather than civil engineering, is offered for taming rivers and for flood protection.

Human Rights and Climate Change

James B. Martin-Schramm

The heavy reliance on fossil fuels (coal, oil and natural gas) together with ecologically damaging land use patterns have produced grave threats to justice, peace and the integrity of creation. The related challenges posed by global warming and climate change are unprecedented in human history. The first half of this article summarizes recent scientific findings about global warming and identifies specific ways climate change imperils human rights around the world. The second half explores two different proposals for securing human rights, which address intra-generational ethical issues related to global climate change.[1]

Climate science

After nearly two decades of intensive study, scientists have reached a much greater consensus about the causes and likely impacts of global climate change. In 1998, the United Nations (UN) established the Intergovernmental Panel on Climate Change (IPCC) to review and assess the most recent scientific, technical and socioeconomic information relevant to climate change. The IPCC has produced reports every five years and issued its Fourth Assessment Report in four installments during 2007. Over 1,200 authors have contributed to the report and their work was reviewed by more than 2,500 scientific experts.[2] Since each report for policy makers is approved line by line in plenary sessions, the IPCC's findings are arguably the least controversial and most accepted assessments of climate change in the scientific community.

---

[1] I have also written a second paper that addresses intergenerational ethical issues related to global climate change and develops ethical criteria to assess the adequacy of various climate policy proposals. The paper is published in the online *Journal of Lutheran Ethics*. See **www.elca.org/What-We-Believe/Social-Issues/Journal-of-Lutheran-Ethics/Issues/April-2009/Assessing-Climate-Policy-Proposals.aspx**. Both papers have been revised and expanded in *Climate Justice: Ethics, Energy, and Public Policy*, to be published by Fortress Press in January 2010.

[2] Intergovernmental Panel of Climate Change, *Fact Sheet for Climate Change 2007*, at **www.ipcc.ch/press/factsheet.htm**, accessed July 2007.

In its Fourth Assessment Report, the IPCC states that it has very high confidence [greater than ninety percent probability] that "the globally averaged net effect of human activities since 1750 has been one of warming."[3] The report demonstrates that, as a result of human activities, global atmospheric concentrations of greenhouse gases such as carbon dioxide, methane and nitrous oxide have substantially increased. For example, the global atmospheric concentration of carbon dioxide, the most important greenhouse gas, has increased from a pre-industrial value of about 280 ppm to 379 ppm in 2005. The growing atmospheric concentration of carbon dioxide "exceeds by far the natural range over the last 650,000 years (180 to 300 ppm) as determined from ice cores."[4]

Direct scientific observations of climate change led the IPCC to declare that warming of the climate system is "unequivocal." It notes that "eleven of the last twelve years (1995–2006) rank among the 12 warmest years in the instrumental record of global surface temperature." The IPCC also identifies "numerous long-term changes in Arctic temperatures and ice, widespread changes in precipitation amounts, ocean salinity, wind patterns and aspects of extreme weather including droughts, heavy precipitation, heat waves and the intensity of tropical cyclones."[5]

These key findings lead the IPCC to the following conclusion: If the world takes a business-as-usual approach and continues a fossil fuel intensive energy path during the twenty-first century, the IPCC projects current concentrations of greenhouse gases could more than quadruple by the year 2100. Under this scenario, global average surface temperature will increase by 4.0°C (7.2°F) by the end of the twenty-first century. Put into perspective, the global-average surface temperature only increased 0.6°C (1.1°F) during the twentieth century.[6] In a report issued after the IPCC released its Fourth Assessment Report, the US Climate Change Science Program warned "[w]e are very likely to experience a faster

---

[3] Intergovernmental Panel on Climate Change, *Fourth Assessment Report: The Physical Science Basis* (Geneva: IPCC Secretariat, February 2007), pp. 2–3, at **www.ipcc.ch/SPM2feb07.pdf**, accessed July 2007.

[4] *Ibid.*

[5] *Ibid.*, pp. 4–6.

[6] *Ibid.* This mean projection is for the fossil fuel-intensive A1F1 scenario, the worst of the six developed by the IPCC. Under this scenario greenhouse gas concentrations are projected to increase from approximately 430 ppm of carbon dioxide equivalent (CO_2e) in 2005 to 1550 ppm CO_2e by 2100. Even under the IPCC's best case scenario, (B1) greenhouse gas concentrations increase to 600 ppm CO_2e by 2100, which they estimate will lead to a warming of 3.2°F by the end of this century—almost three times the rate of warming over the past 100 years.

rate of climate change in the next 100 years than has been seen over the past 10,000 years."[7]

This rapid rate of global warming will raise sea levels, endangering millions living in low-lying areas, despoil freshwater resources for one sixth of the world's population, widen the range of infectious diseases such as malaria, reduce global agricultural production and increase the risk of extinction for twenty to thirty percent of all surveyed plant and animal species.[8] The IPCC emphasizes that poor communities will be "especially vulnerable" to increasing climate change, "in particular those concentrated in high-risk areas" who "have more limited adaptive capacities, and are more dependent on climate-sensitive resources such as local water and food supplies."[9]

Climate change and human rights

Given this warning, it should come as no surprise that some poor and vulnerable communities around the world are beginning to argue that climate change is resulting in violations of their human rights. In December 2005, over sixty Inuit Indians, who live in Arctic regions of the USA and Canada, submitted a petition to the Inter-American Commission on Human Rights. Faced with a rate of warming that is almost twice the pace experienced elsewhere on the planet, the petitioners requested relief "from human rights violations resulting from the impacts of global warming and climate change caused by acts and omissions of the United States."[10]

---

[7] US Climate Change Science Program, *The Effects of Climate Change on Agriculture, Land Resources, Water Resources, and Biodiversity* (September 2007, public review draft), p. 7, at **www.climatescience.gov/Library/sap/sap4-3/public-review-draft/sap4-3prd-all.pdf**, accessed September 2007.

[8] Intergovernmental Panel on Climate Change, *Fourth Assessment Report: Climate Change Impacts, Adaptation, and Vulnerability* (Geneva: IPCC Secretariat, April 2007), p. 8, at **www.ipcc.ch/SPM6avr07.pdf**, accessed July 2007.

[9] *Ibid.*

[10] Inuit 2005, *Petition to the Inter-American Commission on Human Rights Seeking Relief from Violations Resulting from Global Warming Caused by Acts and Omission of the United States*, p. 1, at **www.inuitcircumpolar.com/files/uploads/icc-files/FINALPetitionSummary.pdf**, accessed October 2008. Cited in James Peter Louviere and Donald A. Brown, *The Significance of Understanding Inadequate National Climate Change Programs as Human Rights Violations*, at **http://climateethics.org/?p=39**, accessed October 2008.

In 2008, we observed the sixtieth anniversary of the United Nations Universal Declaration of Human Rights.[11] Drafted originally to protect human dignity after the ravages of World War II, there are several articles in the Universal Declaration which can be applied directly to the perils posed by global climate change. Excerpted below are the most clearly relevant articles in the Declaration:

> **Article 3:** Everyone has the right to life, liberty and security of person.
>
> **Article 7:** All are equal before the law and are entitled without any discrimination to equal protection of the law. All are entitled to equal protection against any discrimination in violation of this Declaration and against any incitement to such discrimination.
>
> **Article 12:** No one shall be subjected to arbitrary interference with his privacy, family, home or correspondence.... Everyone has the right to the protection of the law against such interference or attacks.
>
> **Article 17:** Everyone has the right to own property alone as well as in association with others. No one shall be arbitrarily deprived of his property.
>
> **Article 22:** Everyone, as a member of society, has the right to social security and is entitled to realization, through national effort and international co-operation and in accordance with the organization and resources of each State, of the economic, social and cultural rights indispensable for his dignity and the free development of his personality.
>
> **Article 25:** Everyone has the right to a standard of living adequate for the health and well-being of himself and of his family, including food, clothing, housing and medical care and necessary social services, and the right to security in the event of unemployment, sickness, disability, widowhood, old age or other lack of livelihood in circumstances beyond his control.
>
> **Article 28:** Everyone is entitled to a social and international order in which the rights and freedoms set forth in this Declaration can be fully realized.

Most nations of the world have endorsed the Universal Declaration on Human Rights, and approximately seventy-five percent have ratified other legally binding international laws, such as the International Covenant on Civil and Political Rights (ICCPR) and the International Covenant

---

[11] United Nations, *Universal Declaration of Human Rights*, at **www.un.org/Overview/rights.html**, accessed July 2008.

on Economic, Social, and Cultural Rights (ICESCR).[12] The Inuit Indians appeal to both of these international laws in their petition to the Inter-American Commission on Human Rights. They also appeal to other legally binding agreements. For example, they argue that, as a member of the Organization of American States, the USA must respect their rights under the American Declaration of the Rights and Duties of Man. They also argue that, as a signatory of the UN Framework Convention on Climate Change, the USA has committed to developing and implementing policies aimed at reducing its greenhouse gas emissions.[13] The following quote from the Inuit petition summarizes their argument:

> The impacts of climate change, caused by acts and omissions by the United States, violate the Inuit's fundamental human rights protected by the American Declaration of the Rights and Duties of Man and other international instruments. These include their rights to the benefits of culture, to property, to the preservation of health, life, physical integrity, security, and a means of subsistence, and to residence, movement, and inviolability of the home.[14]

Nearly three years later, the Inuit case remains pending before the Inter-American Commission on Human Rights. While it is clear that human rights enshrined in various laws appear to be jeopardized by global climate change, this does not mean that the Inuits (or others) will prevail in their legal cases. Since climate change is a global phenomenon it is difficult to establish which entities have the jurisdiction and authority to rule in any particular case. In addition, courts find it difficult to assign proportional national or corporate responsibility for the greenhouse gases that have been emitted since the advent of the Industrial Revolution.

Legal matters aside, a strong moral argument can be made that global climate change is causing human rights violations.[15] In 2000, the World Council of Churches (WCC) issued a statement that "equitable rights to the

---

[12] Louviere and Brown, *op. cit.* (note 10).

[13] Inuit 2005, *op. cit.* (note 10), p. 5.

[14] *Ibid.*

[15] See Donald Brown, "The Case for Understanding Inadequate Climate Change Strategies as Human Rights Violations," in Laura Westra, Klaus Bosselmann and Richard Westra (eds), *Reconciling Human Existence with Ecological Integrity* (London: Earthscan Publications, 2008).

atmosphere as a global commons must be the foundation of proposals to address climate change."[16] Michael Northcott, author of *A Moral Climate: The Ethics of Global Warming*, discourages appeals to rights language in Christian responses to climate change.[17] In a recent paper, Northcott argues on pragmatic grounds that "it is not the poor or the weak but the powerful who have most successfully mobilized rights claims in the law courts and economic markets."[18] Moreover, Northcott claims "[t]he assertion of such rights lacks any theological or confessional base in the historic documents of the Christian tradition. More worryingly, rights assertions are a foundational source of violence in the history of the modern world."[19] Northcott argues that throughout Christian history "the ways in which [Christian] communities exercise moral claims on one another have not traditionally been through the language of rights arising from rights or property claims but from obligations recognized in the law of love."[20]

Are appeals to human rights grounded in the norm of justice incompatible with or less important than moral obligations rooted in the norm of love? I think this is a false dichotomy.[21] Rights, in part, specify the content of our moral obligations to others and thus are an invaluable ethical category, particularly for grounding the moral and legal worth of all forms of life. Obligations imply that we owe something to others and, at least, what we owe them are their rights. The problem with a benevolence and duty-based approach is that duties arise from within and rights can only be requested, whereas in a justice and rights-based approach duties arise from without in response to those who are demanding their rights. The fact that rights-based appeals have been abused by the powerful does not change the fact that the concept of rights has been at the heart of successful attempts to achieve greater measures of freedom

---

[16] World Council of Churches, T*he Atmosphere as Global Commons: Responsible Caring and Equitable Sharing: A Justice Statement regarding Climate Change from the World Council of Churches*, at www.wcc-coe.org/wcc/what/jpc/cop6-e.html, accessed October 2008.

[17] Michael Northcott, *A Moral Climate: The Ethics of Global Warming* (Maryknoll: Orbis Books, 2007).

[18] Michael Northcott, "Apocalypse, Accountability, and Climate Change," p. 12, an unpublished paper written for a theological consultation on climate change convened by the Lutheran World Federation in Geneva, Switzerland, 2–4 October 2008.

[19] *Ibid.*, p. 13.

[20] *Ibid.*

[21] Here I draw on the work of James A. Nash who develops a rights-based approach to environmental ethics, in James A. Nash, *Loving Nature: Ecological Integrity and Christian Responsibility* (Nashville: Abingdon Press, 1991).

and equality around the world. While a rights-based approach to ethics can accentuate individualism and undermine community, this would be a distorted understanding of the mutual rights and responsibilities of members in a democratic society. Properly understood, a rights-based approach has the best potential for holding together the twin objectives of protecting individuals and the common good because the purpose of rights is to foster relationality rather than to undermine it.[22]

Rights-based approaches to climate policy

Two organizations have recently proposed a rights-based approach to climate policy in order to protect the interests of poor and vulnerable people around the world. Both address issues related to intra-generational justice. That is, how should burdens associated with addressing climate change be distributed fairly among present generations, and how might benefits be distributed so that vulnerable people are protected from violations of their human rights?

In September 2007, representatives of the Stockholm Environmental Institute and EcoEquity, a think tank devoted to developing "a just and adequate solution to the climate crisis,"[23] published online *The Greenhouse Development Rights Framework: The right to development in a climate constrained world.*[24] Financial support for this project was provided by the large British relief and development organization, Christian Aid,[25] as well as the Heinrich Böll Foundation in Germany, which strives "to promote democracy, civil society, human rights, international understanding and a healthy environment internationally."[26] In an official statement, issued on the tenth anniversary of the Kyoto Protocol, the Executive Committee of the WCC encouraged "further deliberations and negotiations" about "Greenhouse Development Rights"

---

[22] James A. Nash makes this point, in "Human Rights and the Environment: New Challenges for Ethics," in *Theology and Public Policy*, vol. 4, no. 2 (Fall 1992), p. 45.

[23] See **www.ecoequity.org/about.html**.

[24] Paul Baer, Tom Athanasiou, and Sivan Kartha, *The Greenhouse Development Rights Framework. The Right to Development in a Climate Constrained World* (Heinrich Böll Foundation, 2007), at **www.ecoequity.org/docs/TheGDRsFramework.pdf**, accessed October 2008.

[25] See **www.christian-aid.org.uk/**.

[26] See **www.boell.org/overview.asp**.

as the international community develops an agreement to replace the Kyoto Protocol when it expires in 2012.[27]

The authors of *The Greenhouse Development Rights* (GDRs) *Framework* address the impasse which exists between wealthy developed countries and poor developing countries about how to develop an emergency climate stabilization program that will keep global warming under 2°C (3.6°F) by the end of the twenty-first century. While developed nations that ratified the Kyoto Protocol have begun to take some responsibility for reducing greenhouse gas emissions, developing nations such as China, India and Brazil have refused to accept binding greenhouse gas emission reductions that might constrain their ability to develop and improve the standard of living of their citizens. The GDRs framework seeks to overcome this impasse by holding global warming below 2°C "while also safeguarding the right of all people around the world to reach a dignified level of sustainable human development."[28]

The authors emphasize that this right to development belongs to people within nations and not to nations as a whole. Accordingly, the GDRs framework utilizes a "development threshold" based on annual personal income to allocate burden sharing associated with reducing greenhouse gas emissions and increasing sustainable human development. Individuals who fall below this threshold "are not expected to share the burden of mitigating the climate problem," but those above the development threshold "must bear the costs of not only curbing the emissions associated with their own consumption, but also of ensuring that, as those below the threshold rise toward and then above it, they are able to do so along sustainable, low-emission paths.[29]

The authors of the GDRs framework stress, however, that "it should be poor individuals, not poor nations, who are excused from bearing climate-related obligations."[30]

Since international efforts to grapple with climate change have focused on the obligations of nations, the GDRs framework utilizes a "Responsibility and Capacity Index" (RCI) to translate individual responsibility to national

---

[27] See **www.oikoumene.org/en/resources/documents/executive-committee/etchmiadzin-september-2007/28-09-07-statement-on-the-10th-anniversary-of-the-kyoto-protocol.html**. Michael Northcott refers to this statement in his unpublished paper.

[28] *Op. cit.* (note 24), p. 9.

[29] *Ibid.*, p. 16.

[30] *Ibid.*, p. 29.

Human Rights and Climate Change

responsibility. The GDRs framework defines national capacity as the amount by which a country's per capita income exceeds the development threshold. Thus, "the portion of a country's GDP that [falls] below the development threshold would be exempt from being 'taxed' to pay for the global emergency program."[31] The GDRs framework utilizes the polluter-pays principle to define national responsibility on the basis of "cumulative per capita CO_2 emissions from fossil fuel consumption since 1990."[32]

A revised version of the GDRs framework was published in June 2008.[33] In this revision, the development threshold is lowered from USD 9,000 per person (calculated on the basis of purchasing power parity) to USD 7,500 per person, which is equivalent to USD 20 per day. Based on this development threshold, the following graphs indicate how national capacity for responding to the climate and development crises should be fairly allocated among India, China, and the USA.[34] With approximately ninety-five percent of their population living below the development threshold, the national capacity of India is very small compared to the huge capacity of citizens of the USA, which has less than five percent of its citizens living below the development threshold.

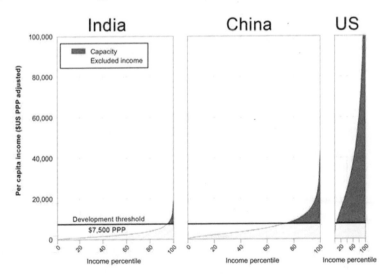

[31] Ibid., p. 31.

[32] Ibid., p. 38.

[33] See **http://gdrights.org/wp-content/uploads/2009/01/gdrs_execsummary.pdf**, accessed June 2009.

[34] Ibid.

The table below calculates national responsibility for cumulative per capita CO_2 emissions since 1990 and combines this with national capacity to arrive at a "Responsibility-Capacity Index" for individual nations and groups of nations.

	Percentage of global total					
	population	income	capacity	Cumulative emissions 1990-2010	responsibility	RCI
United States	4.6	20.7	29.7	23.3	33.9	31.8
EU (27)	7.2	21.6	27.9	15.9	20.5	24.8
United Kingdom	0.9	3.1	4.2	2.1	2.9	3.6
Germany	1.2	4.1	5.6	3.4	4.6	5.2
Japan	1.9	6.1	8.1	4.6	6.2	7.4
Russia	2.0	3.2	2.9	6.3	5.9	3.9
Brazil	2.9	2.8	2.3	1.4	1.2	1.8
China	19.7	12.5	5.9	15.7	7.5	6.6
India	17.2	5.2	0.8	4.2	0.7	0.8
South Africa	0.7	0.7	0.6	1.6	1.4	0.9
LDCs	12.5	1.5	0.1	0.6	0.0	0.1
Annex 1	18.8	57.2	75.1	56.5	73.4	74.6
Non-Annex 1	81.2	42.8	24.9	43.5	26.7	25.4
All high income	15.1	55.2	75.6	50.9	71.4	74.3
All middle Income	46.7	36.4	23.4	42.2	27.8	24.8
All low Income	38.2	8.5	1.0	6.9	0.9	0.9
Global Total	100%	100%	100%	100%	100%	100%

Percentage shares of total global population, income, capacity, cumulative emissions, responsibility, and RCI for selected countries and groups of countries. Based on projected emissions and income through 2010. High, Middle and Low Income categories are based on World Bank definitions.

The authors of the GDRs framework argue that the RCI offers a way for nations fairly to determine their "common but differentiated responsibilities and respective capabilities" under the United Nations Framework Convention on Climate Change.[35] Viewed through the lens of the RCI, the USA shoulders 31.8 percent of total global responsibility, and the nations that comprise the European Union bear 24.8 percent of responsibility, but "the wealthy and consuming classes" in developing nations such as China, India, Brazil and South Africa together bear a little more than ten percent of responsibility as well.[36] The authors argue that the GDRs framework, or a similar approach, is the only realistic way to forge

---

[35] See **http://unfccc.int/resource/docs/convkp/conveng.pdf**, accessed June 2009.

[36] See *op. cit.* (note 33).

international consensus, which will be necessary to hold global warming to 2°C while also securing the right to development for people who are poor.

In July 2008, one month after the revised GDRs framework was published, the Environmental Justice and Climate Change Initiative released *A Climate of Change: African Americans, Global Warming, and a Just Climate Policy for the United States*.[37] This study was a collaborative project between leaders of the environmental justice movement in the USA and Redefining Progress, a think tank dedicated to developing "solutions that ensure a sustainable and equitable world for future generations."[38] The study also frames climate policy within the context of human rights:

> Climate change is not only an issue of the environment; it is also an issue of justice and human rights, one that dangerously intersects race and class. All over the world people of color, Indigenous Peoples and low-income communities bear disproportionate burdens from climate change itself, from ill-designed policies to prevent it, and from side effects of the energy systems that cause it.[39]

Since African Americans are one of the largest minority populations in the USA, the bulk of the study focuses on how African Americans are adversely impacted by various issues related to climate change. Separate chapters address health issues related to warming and natural disasters resulting from increased storm intensity, economic impacts associated with rising energy prices and the cost to maintain US energy supplies and also various dimensions of the persistent phenomenon of institutionalized racism. What follows are key findings excerpted directly from the study's executive summary:

> Global warming amplifies nearly all existing inequalities… .
> Sound global warming policy is also economic and racial justice policy… .
> Climate policies that best serve African Americans and other disproportionately affected communities also best serve global economic and environmental justice… .
> A distinctive African American voice is critical for climate justice.[40]

---

[37] J. Andrew Hoerner and Nia Robinson, *A Climate of Change: African Americans, Global Warming, and a Just Climate Policy for the U.S.*, Environmental Justice and Climate Change Initiative, 2008, at **www.ejcc.org/climateofchange.pdf**, accessed September 2008.

[38] See **www.rprogress.org/about_us/about_us.htm**, accessed June 2009.

[39] *Op. cit.* (note 37), p. 1.

[40] *Ibid.*

All of these findings are addressed to some extent in the sixth chapter of the report, "Elements of a Just Climate Policy." Here the authors assess three different climate policy scenarios and variations within them. Where the GDRs framework focuses more on how the burden of reducing greenhouse gas emissions should be distributed fairly while preserving the right to sustainable development, the Environmental Justice and Climate Change Initiative focuses on how different climate policies impact and can most benefit people who are poor and the victims of racism.

First, they dismiss "phony reductions" achieved through the European Union Emissions Trading System and the Kyoto Protocol's Clean Development Mechanism because both have sanctioned continued investments in fossil fuel infrastructure, ecologically damaging hydroelectric projects and large tree plantations that jeopardize the livelihoods of local communities. They also repudiate carbon offset purchases by rich nations from projects in poor nations because these offsets allow rich nations not to curb their own consumption of fossil fuels. While the authors acknowledge that these policies offer a means to transfer money and technology to developing nations and also to protect the habitat of endangered species, they argue that these policies are currently too flawed to achieve these goods and actually result in "phony reductions."

The Environmental Justice and Climate Change Initiative also repudiates "corporate windfalls" associated with "cap-and-trade" systems where greenhouse gas emission allowances are distributed for free by a government agency to corporate polluters. Instead of requiring polluters to pay for the pollution they emit, the free distribution of allowances rewards the polluters and enables only them to benefit financially from the trade of emission allowances. In this approach to climate policy, "big polluters are treated as though they have a right to pollute and taxpayers and consumers are obligated to bribe them to quit."[41] The authors of the study argue that "cap-and-trade" systems of this sort require the long and complex task of determining baseline emission levels for thousands of polluters and the costly task of verifying emissions reductions at thousands of locations. They also worry that powerful corporations will figure out how to "game" the system both in terms of setting the initial emissions cap and also in terms of how the emission allowances will be allocated. Finally, the report emphasizes that "cap-and-trade" approaches can produce emissions "hot spots" where emissions are concentrated disproportionately in communities of color.[42]

---

[41] *Ibid.*, p. 46.

[42] *Ibid.*, p. 47.

The approach to climate policy most favored by the Environmental Justice Climate Change Initiative is oriented around the polluter-pays principle. Here the authors explore four different options and emphasize that revenue raised from polluters via any of these options needs to be returned to consumers directly and especially to people who are poor.

The first two options focus on governments imposing a fee or a tax on greenhouse gas emissions.[43] Here the goal is to capture environmental costs associated with greenhouse gas emissions in the prices of goods whose consumption results in emissions. Since a fee or a tax sets a fixed price for emissions, this predictability would help consumers and businesses make better long-range decisions about the costs and benefits of less-polluting technologies. In addition, all consumers and businesses are used to the assessment of separate fees or taxes on various kinds of economic activity; thus, a new emissions fee or tax would not be hard to understand.

These two strengths are matched by two significant weaknesses. The first is that few politicians are willing to propose new taxes, and many members of the public perceive new fees simply to be taxes under a different name. The second major weakness is that a fee or tax on greenhouse gas emissions does not guarantee that greenhouse gas emissions will be capped at a certain level, and a fee or tax will likely have to be significant in order for emissions to be reduced. For example, it was not until gasoline prices more than doubled between 2006 and 2008 that drivers in the USA began to reduce their vehicle miles traveled. Sadly, the revenue from this "tax" was sent to oil-exporting nations; it was not captured by the US government to encourage research and development of alternative fuels or to address the regressive impact of these higher energy costs on people who are poor.

The other two options considered by the Environmental Climate Change Initiative revolve around governments establishing a firm cap on greenhouse gas emissions and then selling related emission allowances. In one case, emission allowances are auctioned to polluters by the government; in the other, emission allowances are distributed for free by the government to citizens who then sell them directly to polluters. The study refers to the former as a "cap-and-auction" approach and to the latter as a "cap-and-dividend" approach. In both cases, "collective ownership of the atmospheric commons" is viewed as "a shared birthright" for all people on the planet in contrast to the "cap-and-trade" approach "where

---

[43] *Ibid.*, p. 48.

polluting is a right that belongs to the polluter."[44] In both cases, the more a company pollutes, the more it will have to pay for the necessary emissions allowances. A key weakness associated with the "cap-and-auction" approach is that auction prices could vary significantly in relationship to demand. While the study indicates how this problem could be addressed, corporations would still pass these costs on to consumers and these costs would hit the poor the hardest. A key weakness associated with the "cap-and-dividend" approach is that it would be difficult to empower all citizens equally to sell the emission allowances allocated to them and it would likely result in high administrative expenses.

Regardless which route to climate policy is taken, the Environmental Justice and Climate Change Initiative advocates a "Climate Asset Plan," which is designed not only to provide "climate justice" but also "common justice, justice for all."[45] One way to approach these goals would be to distribute government revenues raised by fee, tax, or auction on an equal per capita basis. The authors of the study note that this approach would yield a disproportionate benefit those who are poor because the payment would represent a larger percentage of income for low-income households. Nevertheless, the authors argue that "a more nuanced approach may allow us to reap even larger benefits for justice, the economy, and African Americans."[46] They endorse distributing sixty-two percent of the emissions revenue on a per capita basis to all citizens, but they propose allocating an additional eighteen percent for energy assistance programs like the Low Income Home Energy Assistance Program, and the remaining twenty percent to promote energy efficiency.

There are certainly many weaknesses associated with the approaches to climate policy advocated by the GDRs framework and the Environmental Justice and Climate Change Initiative. Indeed, many will view them as unrealistic. However, I prefer to view them as moral correctives to policy discussions that too quickly and easily discount the interests and voices of people who are poor and disenfranchised. I agree that legal attempts to seek redress for the violations of human rights caused by global climate change will likely flounder. As a Christian ethicist, however, I think moral appeals to human rights concerns are essential as nation states and the international community come to grips with the unprecedented perils posed by global warming and climate change.

---

[44] *Ibid.*, p. 49.

[45] *Ibid.*, p. 51.

[46] *Ibid.*, p. 50.

Who Dies First? Who is Sacrificed First? Ethical Aspects of Climate Justice

Christoph Stueckelberger

Painful questions

Who dies first as a result of climate change? This question is not as theoretical as it might have seemed twenty years ago. It is an everyday reality for the millions of victims of droughts or storms. It is a painful question for thousands of decision makers who, with limited resources, set priorities for mitigating climate change. Since death as a result of climate warming is not a natural disaster, but caused by human activity, the question becomes even more painful.

The question is not who dies first due to fate, but whom do we sacrifice first because of what we do or fail to do. Do we sacrifice the population of the small islands in the Pacific, whose land disappears? (The president of the Maldives is already looking for another land for his people.) The children in the slums of the megacities who suffer from hunger because of high food prices? Older people in industrialized countries who are infected by new diseases due to climate change? The victims of storms or broken dams?

The notion of humanity as one global village is a rather romantic one. Rather more dramatic is the image of humanity living on the same boat, with the stronger ones starting to push those who are weaker overboard. With regard to global warming, this picture is probably more accurate. We claim to be moral beings who do not want others to die unnecessarily, but it happens every day. The boat is not full; there is space for others. But where are the resources to feed them, cure them and protect them? Who has the will and the power to decide on a fair distribution of the existing resources and the development of new natural, financial, technical, structural and spiritual resources in order to deal with climate change and thus to minimize the number of victims? Who has to pay, and how much, to clean up the damage caused by global warming? Should the polluters pay?

From climate change to climate justice

The central question is no longer whether or not climate change will happen, or whether or not it is caused by human beings. After over twenty years of extensive research, the effects of climate change on the environment, health, migration, politics, economy and culture have become evident, even if further studies on foreseen disasters are still necessary. On the occasion of the twentieth anniversary of the International Panel on Climate Change (IPCC), 31 August 2008, UN Secretary General, Ban Ki-Moon, emphasized the great threat that climate change poses for the UN Millennium Development Goals. The basic ethical question confronting us today is how to distribute the limited resources between three areas—prevention, mitigation and adaptation—in order to minimize the number of victims. Climate change has become a question of global climate justice.[1]

The basic value: justice

The questions, Who dies first? and Who pays how much? are essentially questions of justice. While it is important to include other values such as responsibility and solidarity in ethical reflections about climate change, I shall concentrate on the value of justice/equity.

Throughout history, justice as the fair and just distribution of opportunities, responsibilities and burdens has been a key value in all ethical systems and societies. However, the way in which it is interpreted, or the importance it is given in relation to other values, varies. Climate justice implies just and fair instruments, decisions, actions, sharing of the burden and accountability in order to prevent, mitigate and adapt to climate change. This includes the following dimensions:

Capability related justice means that every person and institution has the duty to contribute to solving problems on the basis of their ability. In relation to climate justice, everyone can and should contribute according to their physical, economic, political, intellectual and spiritual abilities. Any person, institution, company or state who is economically

---

[1] See also *Dossier: Klimawandel und Gerechtigkeit*, at **www.klima-und-gerechtigkeit.de/publikation.html**, accessed June 2009.

Ideals vs. reality

strong has to contribute more to solving the challenges posed by climate change than do those who are economically weak.

Performance related justice means that every institution involved in human activities (such as the production, trade, sale or disposal of a product or service) must be paid on the basis of their performance. In relation to climate justice, an activity that reduces greenhouse gas emissions constitutes a good performance and should be rewarded accordingly.

Needs related justice means that basic human needs and rights (i.e., a living wage, a life in dignity and the right to food and water) should be taken into consideration for every person and institution. In relation to climate justice, every person has the right to that which they need to survive and to adapt to climate change irrespective of their capability and performance.

Distributive justice means that access to resources, goods and services is distributed fairly, taking into account the balance of capability, performance and needs. In relation to climate justice, the financial or other resources needed to decrease the negative effects of climate warming on human life should be distributed primarily according to need, while taking into account performance and capabilities so that the overall disparity between people becomes smaller and not larger.

Justice as equal treatment means that all human beings have the same human rights and the right to equal treatment, irrespective of their capabilities, performance, needs, origin and other factors (such as gender, race, caste, religion). In relation to climate justice, this means that climate related measures for prevention, mitigation and adaptation should ensure the equal treatment of all those affected.

Intergenerational justice means the sustainable use and fair distribution of resources, including between present and future generations. In relation to climate justice, decisions have to respect the needs of future generations who have the same right to equal treatment as do people living today.

Participatory justice means fair and appropriate participation in decision making by all those affected. In relation to climate justice, climate related policy decisions should be arrived at through democratic participation and representation at different levels, from the local to the global.

Procedural justice means calculable, publicly accountable, transparent, corruption free and therefore fair procedures. Decisions related and responding to climate warming (such as access to financial resources, climate related taxes or incentives, media information) have to follow such criteria for procedural justice.

Functional justice means a fair and optimal relation between personal needs and what is possible in terms of institutions, processes and resources. Functional justice concerns where, when and to whom to allocate how much of which resources. In relation to climate justice, this is instrumental with regard to finding fair solutions to the climate challenges.

Punitive justice means that actions violating justice need to be punished through retaliation, deterrence or the overcoming of existing injustices. Where climate justice is violated, measures of punitive justice need to be as severe as for other crimes.

Transitional justice means that in transformational societies, where ordinary, regular institutions and procedures may not exist or be under reconstruction, provisional justice is needed. In relation to climate justice, in emergency situations, accelerated procedures for decisions to provide aid and exceptional instruments such as amnesty (not impunity) may be required and ethically justified.

Restorative justice means that justice between perpetrators and victims needs to be restored through measures of compensation, reparation and/or reconciliation. Climate injustice occurs daily because those who suffer the most from the negative impact of climate change are not those who cause it. Polluters, especially in industrialized countries, need to take bold and decisive measures to restore justice.

Transformative justice is a creative, ongoing process that goes beyond punitive or restorative justice to transform and renew reality toward greater justice. Climate justice does not mean taking isolated decisions or actions, but is an holistic process of transformation of societies in their relations, the use of natural resources, distribution of goods and services and sustainable policies.

On-time justice means that decisions and actions for the sake of justice need to be taken at the right time (in German: *zeitgerechte Entscheide, Zeit-Gerechtigkeit*, in Greek: *kairos*, the right moment). If this occurs too late, the patient or victim dies unjustly and it is difficult to restore justice. This time factor is critical for the sake of climate justice, to prevent further victims of climate change.

Some of these fourteen aspects of climate justice are in tension with others and it is difficult to implement all of them at the same time. Yet, this overall list makes us aware that climate justice is more than just a slogan. It is a fundamental value with a specific, challenging content.

Some ethical guidelines for climate justice

On the basis of the above, how can ethical guidelines be developed for investing in and distributing the limited resources needed for climate change prevention, mitigation and adaptation in order to minimize the number of victims?

Guidelines only indicate a general direction. Respective situations have to be analyzed and the guidelines interpreted and adapted according to concrete situations. In many situations, conflicts between values exist and priorities need to be set. The following guidelines can help to prioritize and solve conflicts between different values, for the sake of sharing benefits, burdens, power and space.[2]

Resources. First, efforts must be made to increase the overall resources available. This, in turn, reflects how priorities are set and the ethical values undergirding them. High priority must be given to the climate change challenge, since it affects all of humankind, including future generations and the rest of creation. The resources are financial–from governmental, private and not-for-profit sources–but also include other human, organizational, intellectual and religious resources. A global climate fund can be one important instrument for substantially increasing financial resources. Taxing CO_2 and other emissions, in order to reduce them, can generate funds for mitigation and adaptation.

---

[2] See Christoph Stueckelberger, *Umwelt und Entwicklung. Eine sozialethische Orientierung* (Stuttgart: Kohlhammer Verlag, 1997), pp. 30–33. Case study on the UNCED climate convention. Chinese edition, Bejing, October 2008.

Prevention aims at taking action that will lessen the future damage of climate change. It seeks to minimize the number of victims, both today and in the future, for the sake of intergenerational justice. Ethically, prevention and mitigation have higher priority than does adaption. Prevention is also more efficient because costs are lower than for adaption.

Mitigation aims at reducing the present negative effects of climate change, to slow it down and reduce the number of victims.

Adaptation accepts that climate change is occurring, and thus aims to adapt places, people, lifestyles, migration patterns, production, technologies, spirituality, ways of managing conflicts, etc. accordingly. Justice is reinterpreted in terms of the ability to adapt to the new challenges; adaptability becomes a basic human need.

Measures for ***prevention, mitigation and adaption*** are often interlinked. Planting trees and reducing fossil energy consumption are preventive measures that help mitigation and adaption. Resources invested in adapting to emergencies should not be at the expense of long-term preventive measures.

The polluter-pays principle means that polluters should pay for the damages caused by their behavior/action. This broadly accepted principle in environmental ethics is put into practice, for example, in waste management, but is not yet widely implemented in relation to climate change. Climate related taxes on fossil energy have to be increased significantly if this principle is to be taken seriously.

The capability to contribute principle means that responsibility is not only in terms of who causes the damage but who has the ability to contribute to solutions. This can be financially, as well as in terms of scientific research and political, social or spiritual support.

The ***Responsibility and Capability Index (RCI)***[3] is a helpful measuring instrument, which corresponds to the polluter-pays principle as well as the capability to contribute principle. The RCI combines the

---

[3] It is part of the Greenhouse Development Rights (GDRs), mainly developed in Great Britain by development agencies such as Christian Aid and supported by Bread for All, the Swiss Catholic Lenten Fund, etc., see **www.ecoequity.org/GDRs**, accessed June 2009.

cumulated CO_2 emissions of a country and its purchasing power parity and the distribution of wealth. The industrialized countries therefore have the biggest share to pay, but developing and transition countries with purchasing power and a wealthy élite are called to contribute accordingly. This country index shows a way of sharing climate change burdens. It is a serious ethical effort to make climate justice measurable and politically operational.

The combination of positive and negative sanctions. Justice in general and climate justice in particular can be strengthened by positive sanctions (such as incentives, repayments, awards, access to services, etc.) and negative sanctions (taxes, punitive measures, court rulings). Positive sanctions are ethically preferable, because they encourage right behavior. Nonetheless, often negative sanctions to establish punitive justice are also necessary, especially to change the behavior of those who do not respond to positive sanctions. Both mechanisms consider emitting CO_2 and other greenhouse gases to be unethical.

Efficiency and transparency are key factors in the good stewardship of limited resources. The efficient use of resources (energy, capital, organizational structures, intellectual creativity, etc.) can reduce costs, help more people and save more lives. This expresses responsibility and sustainability for future generations. Transparency supports the efficient use of limited resources by reducing corruption, abuse and wrong investments. Transparency and efficiency are important aspects of procedural justice.

Market related instruments. International free market mechanisms contribute substantially to general economic growth and global interaction and peace. Programs such as "Financing for Climate—Innovative Solutions and New Markets"[4] seek to persuade the private sector to see climate change as a business opportunity. Without private investment, climate related funding will never be sufficient.

Market related instruments alone cannot solve the major problems of poverty eradication, fair distribution and climate stabilization. Instead, they aggravate these problems. Today, climate change can be seen as the biggest

---

[4] Title of a conference of the Swiss State Secretariat for Economic Affairs, the International Finance Corporation (IFC) and Swiss Re, 11–12 September 2008, Zurich.

market failure in human history.⁵ Can the market then solve it? From an ethical point of view, the answer lies in the criteria for climate justice: when and where market mechanisms strengthen the above aspects of climate justice, they can be supported. When and where they weaken or violate climate justice, the market has to be replaced or accompanied by binding corrective instruments, such as social and environmental laws regulating markets.

There are many examples of how companies enhance their profit and reputation through activities to reduce climate change emissions.⁶ Trading CO_2 certificates is one form of positive and negative sanctions, based on market mechanisms. Insofar as it really contributes to a reduction in worldwide CO_2 emissions and climate justice, this is ethically positive. If, however, it is abused to circumvent legal restrictions to avoid reorienting activities toward climate justice, or to become morally "purer," it has to be rejected ethically.

Care for the weakest: "Solidarity with the Victims of Climate Change" was the title of an important statement by the World Council of Churches (WCC) in 2002.⁷ To care for the most vulnerable groups in society in cases of emergency corresponds to the human ethos in many cultures, especially in the Judeo-Christian value system. "The option for the poor" is expression of this. To care for the victims and the weakest among them is a guideline for decision making related to climate justice.

This also opens up many questions such as, Who are the victims? Who are the weakest among them? The children and women on Fiji who lose their agricultural land or the elderly people in a suburb of Paris who die from extreme heat? Does justice as equal treatment not require that all people in danger receive the same treatment? Since there are many more people needing support for mitigation and adaptation than resources available, what other criteria are needed? What political calculations and economic conditions come into play? Under what conditions will the empowerment of those who are most vulnerable lead to efficient solutions and the use of scarce resources?

---

⁵ This is the view of the *Stern Review on the Economics of Climate Change*, HM Treasury, UK (2006), at **www.hm-treasury.gov.uk/stern_review_report.htm**.

⁶ Swiss Re (2008a): *Pioneering Climate Solutions*, Zurich. Swiss Re (2008b): *Corporate Responsibility Report. Committed to Sustainable Value Creation*, Zurich.

⁷ *Solidarity with Victims of Climate Change*: Reflections on the World Council of Churches' Response to Climate Change, at **www.oikoumene.org/fileadmin/files/wcc-main/documents/p3/Solidarity_with_victims_of_climate_change.pdf,** pp. 25f., accessed June 2009.

The first step is the honest recognition that often support is not extended to the weakest, even when this is one ethical criterion. A second step is to look for preferential rules when different aspects of justice compete with one another. To give priority to the weakest may meet the criteria of needs-based justice, but it may not be the most ethical if it ignores other aspects of justice. In some cases, more lives may be saved if priority is given to people who are able to use limited resources wisely and efficiently in order to support others to survive. The general ethical preference is that priority be given to the weakest, but consideration also needs to be given to the total number of lives that will be saved.

Institutionalized solidarity: While solidarity involves voluntary care for others, it also needs to be implemented through binding institutionalized measures. New forms of climate related insurance to cover droughts or floods are an example of institutionalized solidarity, combined with arrangements for microcredit.[8]

Urgent legislation. The speed of climate change shows that binding measures for prevention, mitigation and adaptation have to be implemented at a much faster pace than they have been thus far. This has been so slow because of the lack of political will and the slowness of democratic decision-making processes. For more than ten years, the Swiss parliament has been seeking a compromise legislation for CO_2 emissions. "On time justice" is crucial in order to reduce the number of victims. Urgent governmental legislation may be needed and ethically justified even if it limits participatory justice. In emergencies, the rights to food, water and survival have priority over the right to participate in decision making, or in avoiding timely decisions.

Lessons learned and not yet learned

The World Climate Conference in Toronto, June 1988, asked for a twenty percent reduction of CO_2 emissions by 2005 and fifty percent by 2035. During a global conference in Washington in October of the same year, a global, multistakeholder "International Network against Climate Change" was established, at which I participated as the only representative of the Confer-

---

[8] See the recommendations of the round table on, "Are the Right Risks Insured?," at the Global Humanitarian Forum Geneva, 24 June 2008, at **www.ghf-ge.org**, accessed June 2009.

ence of European Churches (CEC). Most of today's facts had already been on the table: prognoses about rising sea levels, changes in food production, droughts, storms and the emergence and reemergence of diseases. On this basis, I wrote in an article some twenty years ago that rivers will start to become salty because of rising sea levels, the supply of drinking water will be threatened, new diseases will emerge, food production will be reduced and environmental migration will increase.[9] The only thing I did not expect was that many of these prognoses would already be a reality in 2009. In 1989, the Swiss Ecumenical Association Church and Environment, of which I was the founding president, started the first climate campaign in Switzerland, which called for a new lifestyle and politics with an annual reduction of energy consumption of two percent in order to implement the Toronto goals.[10] Many regarded this as unnecessary and idealist but, today, twenty years later, the IPCC's figures on hunger victims and drought are even more drastic.[11]

To summarize the development of the climate change positions over the past twenty years:

Scientists have been the early warning and alert system. They were among the first to analyze what has been occurring. The fact that from early on they have coordinated their views worldwide has helped to raise awareness. After twenty-five years of continual research, a global consensus could be reached by the IPCC.

Since the beginning, **politicians** have been divided, often defending the interests of their countries and economies. While some have underlined the urgency of putting in place (common, global) action plans and actions, others have denied the facts. For a long time, the developing countries have held especially the industrialized countries responsible. After twenty years, a consensus regarding global warming, the urgency of addressing it and the economic implications has been reached, but the political will for far-reaching actions is still lagging behind. Much more money is still being spent on regional and local wars than on binding measures to fight the common global "enemy" of global warming.

---

[9] Christoph, Stueckelberger, "Die Treibhauswelt im Jahr 2035. Statt Wintertourismus holländische Flüchtlinge in den Alpen?," in *Kirchenbote für den Kanton Zürich*, no. 20 (October 1988), p. 2.

[10] The title of the campaign was: "Die Haut der Erde retten. Lebt Jahr für Jahr mit 2% weniger Energie." ("Sauvez la peau de la terre. Vivez chaque année avec 2% moins d'énergie").

[11] IPCC, *Climate Change 2007* (Geneva, 2007), at **www.ipcc.ch**, accessed July 2009.

Over the past twenty years, **churches and religious communities** worldwide have been actively promoting reflection, worship and actions on climate change. Between 1990 and 2008, the WCC alone published over twenty booklets, study papers, reports and statements for UN conferences on climate change.[12] In 1990, the first global statement called "to resist globally the causes and to deal with the consequences of atmospheric destruction."[13] The WCC's position papers combined theological and ethical reflection with practical and political recommendations, looking at climate change also as a "spiritual challenge"[14] and affirming that a whole "vision of society is implied."[15] The call for a change in personal lifestyles was combined with the commitment to strengthen common political solutions through the UN system.

Nonetheless, the churches have often not been heard in the media, the parishes and at secular conferences. Decentralized structures and a lack of binding decision-making structures often make it difficult for churches to implement what should be done. Churches have underestimated the influence of technology, marketing and price on human behavior. The 2007 Stern Report, showing the economic effects of climate change, revealed that economics still has the greatest impact in terms of changing behavior. Churches continue to provide input at the global UN level but should also intensify the dialogue with the private sector on climate change.

Other religious communities have started to deal with climate change, especially Muslims and Buddhists, but not with the same systematic and long-term approach as the WCC. Caring for creation is the common ground for interreligious spiritual reflection.[16]

---

[12] See the selected bibliography of the WCC Programme on Climate Change, at **www.wcc-coe.org/wcc/what/jpc/earthdocs.html**, accessed June 2009.

[13] World Council of Churches, *Now is the Time. Final Document and Other Texts of the World Convocation on Justice, Peace and the Integrity of Creation, Seoul 5–12 March 1990* (Geneva: WCC, 1990), p. 31.

[14] The World Council of Churches, *Solidarity with Victims of Climate Change: Reflections on the World Council of Churches' Response to Climate Change*, at **www.oikoumene.org/fileadmin/files/wcc-main/documents/p3/Solidarity_with_victims_of_climate_change.pdf**, pp. 25f., accessed June 2009.

[15] *Ibid.*, pp. 13–15.

[16] See Satria Candao, "Islamic Wisdom and Response to Climate Change," in Carlos B. Mendoza (ed.), *Search for Better Tomorrow. A Consultation on Earth is our Home: A Religious Response to Climate Change in Asia, July 2000* (Bangalore, 2002). See also the World Council of Churches' Interreligious Conference on Climate Change, November 2008, at **www.oikoumene.org/fileadmin/files/wcc-main/2008pdfs/WCC_ClimateChange_BackgroundInfo2008.pdf**, accessed June 2009.

For a long time, *faith based development agencies* have been key actors in raising awareness on energy issues and climate change. Among the Christian agencies, especially Christian Aid in Britain,[17] Bread for the World[18] and EED in Germany, Bread for All and the Swiss Catholic Lenten Fund have prepared important campaign material.[19] This is increasingly being done in cooperation with other actors such as Oxfam.[20] A broad global coalition on climate justice was established.

The private sector is very diverse and cannot be seen as one block. International companies, especially in the insurance sector, have taken the lead within the private sector. Not theoretical or ethical reflections, but precise analyses and long-term practical implementation and financial commitment make them credible.[21]

Threat or opportunity?

Based on the experiences that crises can also lead to renewal, the private sector is beginning to speak about climate change not only as a threat, but also an opportunity. It can motivate people, institutions and companies to find solutions and at least to mitigate the problem. Companies interpret opportunities as business opportunities to sell new products and services. The global Carbon Disclosure Project (CDP), a network of 315 institutional investors, representing assets of USD 41,000 billion, assesses climate related risks and opportunities of the companies in which they invest. This is an example of using economic mechanisms to redirect investments and activities in a climate friendly direction. A global news service, specialized in climate change information for busi-

---

[17] See the material on, **www.christianaid.org**, accessed June 2009.

[18] Diakonie Katastrophenhilfe, Brot für die Welt, Germanwatch, *Climate Change, Food Security and the Right to Adequate Food* (Stuttgart, October 2008), at **www.germanwatch.org/klima/climfood.pdf**, accessed June 2009.

[19] Campaign "Gerechtigkeit im Klimawandel," 2009, at **www.oekumenischekampagne.ch/cms/index.php?id=10**, accessed June 2009.

[20] *Klima der Gerechtigkeit. Entwicklungspolitische Klimaplattform der Kirchen, Entwicklungsdienste und Missionswerke* [Climate of justice. A platform for climate and development promoted by churches, mission agencies and development services], Bielefeld 2009, at **www.ekvw.de/fileadmin/sites/ekvw/Dokumente/texte/Klima_der_Gerechtigkeit_screen.pdf**, accessed June 2009.

[21] *Op. cit.* (note 6).

ness opportunities, was started in 2008.[22] The limitation is that is sees opportunities only from the perspective of companies, while climate change can also be an opportunity for other sectors of society.

Climate change shows more clearly than ever before to what extent human beings depend on one another. It is this concrete experience of global interdependence and interconnectedness that provides an opportunity for increased solidarity and mutual responsibility. Climate change shows that isolated actions are not enough, but that multilateral coordinating global structures and mechanisms are needed. Unilateral, bilateral or autonomous actions alone cannot bring about mitigation and adaptation.

A new lifestyle and society,[23] not based on fossil energy and carbon emission, are possible. To leave existing lifestyles and to look for new ones is an inner journey, involving psychological and spiritual processes of mourning and reorientation.

The crisis of climate change is an opportunity for increased inter-religious cooperation. Not only all sectors of societies, but also all religions are challenged to find answers to the burning spiritual questions climate change has given rise to. Climate change can lead to a deepening and renewal of faith, giving space for mourning and the power of hope expressed in new and renewed confessions of faith.[24]

Spiritual responses: too late or is there hope?

These disastrous prognoses can lead to a sense of resignation. Believers struggle with God's promise to Noah that "never again shall all flesh be cut off by the waters of a flood, and never again shall there be a flood to destroy the earth" (Gen 9:11). There is increasing theological reflection at different levels. "Signs of Peril, Test of Faith" was the subtitle of a 1994 WCC study paper on climate change.[25] Regional responses for

---

[22] Newsletter can be ordered, at **www.climatechangecorp.com**, accessed June 2009.

[23] See Schweizerischer Evangelischer Kirchenbund, *Energieethik*, SEK Position 10 (Bern, 2008).

[24] Churches formulated confessions related to globalization and economic injustice, e.g., the World Alliance of Reformed Churches with the "Accra Confession," 2004. Others ask whether climate change will be part of new confessions: "Gehört auch der Klimawandel in ein neues Bekenntnis?"in *Reformierte Presse* Nr. 30/31 (25 July 2008), pp. 6–7.

[25] World Council of Churches, *Accelerated Climate Change. Sign of Peril, Test of Faith*, approved by the Central Committee of the World Council of Churches (Geneva: World Council of Churches, 1994).

example from Africa[26] or Asia[27] and a collection of contextual responses have been a part of a 2008 process in the Lutheran World Federation (LWF). Crucial questions on climate change include:

- Is it too late or is there hope?

- How are God's promises, not to destroy the earth again but to save lives, to be understood?

- Where are God and the cosmic Christ (Col 1)? Where is God's spirit in climate change? What is God's action?

- What is the role of human responses to God's action? Can we and must we as human beings save the world?

- Who is guilty and how do we deal with guilt? What do forgiveness and reconciliation mean in this context?

- How can we live out our responsibility?

While these questions may seem difficult and even discouraging, how we answer them will either motivate us to or discourage us from taking action. Three types of answers are ethically unacceptable: cynicism and fatalism violate the dignity of victims and do not take their suffering seriously, and fundamentalism seeks to find answers in the past without adapting them to the complex reality of today's climate change. The differentiated answers from a Christian perspective can empower and encourage us to take decisive action.

First, we have to recognize that climate change is an unprecedented global challenge in the history of humankind. Catastrophes such as wars, droughts, floods, accidents or sickness and other experiences of disaster provoke similar questions of faith.

It is very late, but not too late.[28] When we look closely at the data we find that while the pessimists are right—it is too late and temperature

---

[26] Ernst M. Conradie, *The Church and Climate Change*, manuscript (South Africa, 2008); Jesse Mugambi, *Environment and Spirituality: Theological Considerations*, manuscript (Nairobi), p. 8.

[27] Mendoza, *op. cit.* (note 16).

[28] Wolfgang Huber, *Es ist nicht zu spät für eine Antwort auf den Klimawandel. Ein Appell des Ratsvorsitzenden der Evangelischen Kirche in Deutschland*, unpublished manuscript (Berlin, 2007).

will rise by more than the two degrees that are seen as the maximum increase if we want to avoid great catastrophes—the optimists are also right: if we make the efforts required, we can make it. Faith has a different perspective: hope is neither captive to pessimistic nor optimistic interpretations, but it is oriented toward what is promised. I think of myself as a pessimist full of hope: a pessimist when I look at the world, but full of hope when I look at God's promise.

While God promises a life in dignity on earth for all, human and non-human, God did not promise a certain lifestyle. Adaptation is part of life. God promises to accompany us on this journey but did not promise to maintain the world and nature in its original form. Creation is an ongoing process of transformation. Humankind is called to continue this journey of nature and culture and constant change, always seeking orientation through constant dialogue with God. The covenant with Noah was not forever. God's history with humankind shows that the covenant was repeatedly broken by human beings and renewed by God time and again: with Abraham (Gen 17:2), with Jeremiah (Jer 31:31) until the new covenant in Jesus Christ (Mt 26:28). Therefore, the promise was not given once and for all, but has to be renewed with each person and each generation looking for this promise in faith and asking God for this covenant. The renewal of the promise is the result of the relationship between God and believers. The content of God's promise is that God is willing to renew the covenant time and again—if we accept it. That is the source of Christian hope. Human engagement for prevention, mitigation and adaptation to climate change is the test of this hope.[29]

God's promise is empty without this relationship to humankind. Because God is love, God does not do this apart from human beings and all living creatures. God became incarnate in this world in such a way that God bound Godself to this creation.

God's providence means that God cares for and suffers with all living beings. But it is not an automatic, "natural" mechanism or guarantee to save lives. Providence as creation and history is an ongoing, living process. God is the living "motor," "driver," "communicator" and "partner" of human beings in it. As Trinity, God acts as constant Creator, Redeemer and Renewer.

---

[29] As also expressed by the German Catholic Bishops Conference in its statement on climate justice, Die deutschen Bischöfe, Kommission für gesellschaftliche und soziale Fragen, "Der Klimawandel: Brennpunkt globaler, intergenerationeller und ökologischer Gerechtigkeit. Ein Expertentext zur Herausforderung des globalen Klimawandels," in *Kommission Weltkirche*, no. 29 (2006), p. 70.

"Who is sacrificed first?" is not a cynical question but, unfortunately, a daily reality. Christian faith loudly protests against letting people die and be sacrificed. From a faith perspective, human beings are called to do all they can to avoid it. The reason lies in the very heart of the Christian faith: Jesus Christ resisted all evil and answered it with love to a point where he gave his life as a sacrifice "once and for all." No human life has to be and shall be sacrificed after Jesus Christ's last sacrifice.

The prophets of the Old and New Testaments are a rich source for learning how to deal with drastic threats in a given time. The prophets of doom interpret the "signs of the time" (such as war, natural or human-caused disasters, the collapse of power structures) as an expression that human beings did not listen to God's wisdom and will and therefore broke the covenant with God. According to the Prophet Ezekiel (Ezek 26–28), in around 500 BC, the ancient global trading system and immense wealth of the trade town of Tyre collapsed because the king of Tyre had exploited and exported the population of whole villages as slaves, and put himself in God's place (Ezek 28:1). The prophet interpreted the collapse as a result of human arrogance and superciliousness. He called for *metanoia*, a fundamental change in orientation and lifestyle, in order to overcome this catastrophe. The prophet of doom becomes a prophet of hope because he analyzes the reasons for the disaster and shows a way out. The crisis was a threat which turned into an opportunity for reorientation and more humane behavior. Encouraging people to undertake this reorientation is the prophetic role of the churches and other religions.

At the same time, the Christian faith underlines that we do not have to save the world and bear the world on our shoulders alone. Capability related justice and responsibility mean carrying what we can carry, knowing that God supports, accompanies and does not ask us to bear more than we can carry. Sharing the burden leads toward climate justice.

An LWF Climate Change Encounter in India[1]

From 16–20 April 2009, around twenty-four persons from India and other parts of the world met in the coastal community of Puri, in Orissa state, India, to witness firsthand and better to understand dramatic examples of climate change in that area and to reflect on how this relates to what is occurring in other parts of the world. This event was organized by the Department for Theology and Studies of the Lutheran World Federation (LWF), as part of the overall LWF strategy related to climate change, and in cooperation with the United Evangelical Lutheran Church of India (UELCI) and Lutheran World Service India (LWSI).

Participants came from LWF member churches in India, Denmark, Germany, Sweden, Australia, Indonesia and the USA, from LWF-related World Service programs in India, Bangladesh and Tanzania, as well as other Christian denominations and other faiths. They included pastors, theologians, biblical scholars, ethicists, communicators, anthropologists, church staff, advocates, students and specialists working on adaptation and mitigation measures related to climate change.

Those from outside India first went to Calcutta, an intense sprawling metropolis to which millions of people have migrated from other places, such that today, only thirty-seven percent speak the local language, Bengali. Calcutta is becoming a center for many "climate refugees," who migrate there from areas especially impacted by climate change. Through the Urban Development Program of LWSI, participants encountered and spoke with those living in a settled community along a railroad track behind a massive garbage dump, and in another community beneath a railroad track embankment. In both cases, residents expressed how they are being empowered to take responsibility for their communities and lives.

During our time together in Puri, we analyzed the causes and effects of climate change around the world, reflected from biblical and other faith perspectives, and spent two very hot days of exposure visits to rural coastal com-

---

[1] Participants included: Arul Aram, B.N. Biswal, Karen L. Bloomquist, Sagarika Chetty, Norman Habel, Keld Balmer Hansen, Anupama Hial, Anam Chardra Khosla, Bonar Lumbantobing, Chandran Paul Martin, Peter Matthews, Kishore Kumar Nag, Belinda Praisy, M. G. Neogi, Sofia Oreland, Barbara Rossing, Peniel Rufus, Richard Sarker, Emmanuel Shangweli, Anja Stuckenberger, Gnana Theophilus, Wesley Vinod, Annie Watson.

munities dramatically affected by climate change. These were communities in which the Rural Development Project of LWSI is working to educate, empower people and support local initiatives (self-help groups, disaster management and village development committees) in the face of these changes.

A number of the participants and most living in the villages were Indigenous Peoples (Dalit, tribal/Adivasis). Creative worship was led by recent graduates of Gurukul Lutheran Theological College, with Indigenous music, symbols and stories underlining the spiritual meaning and significance of what we saw and heard during the visits.

In the six rural communities we visited, we were warmly welcomed by the hundreds of people residing in these villages with Indigenous songs, rituals and flowers. We heard testimony from and dialogued and interacted with a large number of persons, whose entire lives, meaning and future are deeply affected by climate change. They intimately know its realities and causes and were eager to share their knowledge and experiences with us.

The people we met invited us to witness and to be witnesses for and with them, which is the purpose of the following communiqué. They yearn to make global connections and experience solidarity with others in what they have experienced and in the hopeful actions they are taking for their future. The following communiqué conveys what we witnessed.

Asha:[2] witnessing to hope amid rising waters

In many parts of the world, climate change remains something "out there" in the future, but for these villagers it is no longer "out there." With every surge of the insatiable sea, climate change becomes all the more local, to the extent that one day their village will be "in there," swallowed by the waters. Here the disturbing effects of climate change are not just predicted to occur in the future, but are undeniably present today.

Climate change scientists may be among today's prophets, but the people and land in these communities bear vivid testimony to the actual pain and displacement involved. Whole villages have been destroyed by the sea. More frequent and extended periods of rain, along with more intense storms, flood their land and homes far more often than in the past. People here live with this ever-present reality, which haunts their memories, motivates what they are doing in the present and shapes their hopes for the future.

---

[2] The word for "hope" in several Indian languages.

What we saw and experienced

As we looked out at the rising waters of the Bay of Bengal, and heard the waves pounding onto the shore each morning, we saw large expanses of seawater and shrinking sandbars, covering large areas where a few years ago, hundreds of houses had stood, along with land for growing crops. The sea is coming much closer than it has been before, whether gradually or through sudden cyclones: fifteen years ago, people went out twenty kilometers to the sea, but now it is only half a kilometer away and coming closer every day.

> We stood on the shore and looked out to the sea of the Bay of Bengal. Beside us on the beach were the shells of large turtles whose habitat had been destroyed. A few hundred yards out to sea was a sandbar shining golden in the sun, which we were told would be submerged in another two months. Only a few years ago that shining sandbar was a lively fishing village. Now there is nothing—no houses, no animals, no trees. The village had been swallowed by the sea. Like several other villages along the coast of Orissa on the Bay of Bengal, the sea had inundated the coastline and swallowed all before it. Fifteen years ago, said one fisherman, the shoreline was five kilometers away. Now it was only a few hundred meters away.
>
> Nearby, casuarina trees, provided through LWSI, are helping to hold back the further encroachment of the sea into the land, as well as protecting the Hindu temple that has replaced the previous temple that disappeared into the sea.

What we heard

The state of Orissa is the poorest of the eleven major Indian states; nearly half of the rural population live below the poverty line. The state has the highest infant mortality rate and fifty percent of the malaria related deaths in India. Tropical storms that form over the Bay of Bengal have made this one of the most disaster prone areas of the world. We heard and experienced how Orissa is reeling under climatic chaos ranging from heat waves to cyclones, from droughts to floods. In the last four years alone, such calamities have claimed more than 30,000 lives.

People's lives and identities are deeply connected with the land and the water. For their livelihood, they fish the waters and farm the land. Their intrinsic connection with the land and the water is being impacted dramatically: climate change is contributing significantly to there being too much water, with the strong winds (or cyclones) increasingly flooding the land, destroying life and livelihoods. Their water for drinking and irrigation becomes salty from the invading sea, or polluted because of the lack of adequate sewage. Further inland, periods of too little water (drought) are the problem. When asked why all of this is occurring, even the young children could readily describe why the climate is changing.

We felt firsthand the pattern of excessive, rising and moist heat that is seriously impacting health and life conditions for human beings, animals and plants. Villagers reported dramatic rises in waterborne diseases such as sunstroke, skin diseases, malaria, dengue, hepatitis B, arthritis, etc. Plants and fish are smaller and more susceptible to disease.

Furthermore,

- Farming, fishing and drinking water has become polluted and salinized: "What was life-giving water now is killing us."

- The seasons are changing—rains come later and heavier, and the summer heat is longer and more intense.

- Residents spoke of land around them that used to be covered with jungles, providing habitat for tigers, but on which only a few protected trees now stand.

- A river that used to flood every five to seven years now does so every year, wiping out the crops on which they depend for their livelihood. The people conveyed a calm sense of inevitability, "the water will come and wash everything away again."

- When asked to reflect on what they had experienced, they replied, "The pain has rooted in our hearts—every day we live with the pain. It has become part of our lives."

- "Fifteen years from now this village will not exist. But why would we move? This is where our life and livelihood are." A young man admitted that he might have a better life in the city, "but what then will happen to this community?" Again and again, the refrain was heard, "We will not leave the land!"

When a super cyclone hit the coastal village of Natara, the storm winds ripped the roofs from the houses. The farm animals fled and drowned in the sea. Amid the mayhem, one house with a metal roof remained standing. The people of the village huddled in this house and made a joint decision: "We will not flee. If we live, we will live together. If we die, we will die together." For five days they survived under that roof—no food except for a few fallen coconuts that provided some sustenance for the children. When the storm subsided, they gathered what food they had and continued to eat their meals in common. "The cyclone has brought us together as a community." Among other measures, they are now building a dyke to keep the encroaching water out of their village.

The locals have no doubt that the rising sea levels are due to climate change. The devastation of their village is evidence they cannot ignore. We have heard about the probability that islands in the Pacific, for example, will one day be inundated by rising seas. But, for the villagers on the Orissa coast that day has come.

Yet we were encouraged

The strong bonds of care and communal sharing in these Hindu and tribal villages are reminiscent of the early Christian communities in which they shared all things (Acts 2:44–45), as well as the story of the feeding of the multitude (Mt 14:13–21): through sharing, there were bread and fish for all. Might communities such as these in India inspire all of us to do more of this today? People are yearning for different ways of living that will not make them so sick.

Climate change is dramatically affecting the physical environments of these fishing and farming villages. Changes in the natural environment caused, for example, by climate change, deforestation and pollution not only affect the ecosystem and economy, but also Indigenous ways of life. People spoke about the current changes from practical but also spiritual perspectives; these are deeply intertwined.

For the people in these villages, religious beliefs and practices are intimately connected to the natural domain. The elements of nature, such as the sea, trees, land and animals are thought of as divine beings. They are revered in everyday practice and celebration. Trees set aside in the community are shrines for deities.

Among other factors, it was through these spiritual perspectives that people interpreted their experiences of natural disasters and furthered social, economic, psychological and ecological processes of adaptation and restoration. Although they sensed that the gods might be punishing them because of what they have done to nature, they view themselves as responsible for setting things right again. Strong bodied fishermen cast their nets into the sea of their livelihood, with the hope that the goddess of the sea will not destroy again.

Based on these observations, we recommend that projects and initiatives for relief, development and empowerment be sensitive to and include local spiritualities and ways of life through which people relate to nature and to each other. Attention to this is crucial if responses to climate change are to be effective and sustainable.

We were continually amazed by the enduring hope of the people, and their resolve and commitment to act, adapt to and change the situations they face:

- Their trust in their gods to protect and guide them: "We believe that our village will disappear in the next ten years, but we also believe that the gods will take care of us."

- In observing one religious festival, for three days the land is not touched, but is "treated like a woman undergoing menstruation." During other festivals also, earth is given rest. There are connections here with the Sabbath traditions in the Bible.

We were impressed by their initiatives to adapt and take preventative measures by:

- Continually planting more trees

- Educating the children

- Promoting traditional food, well-being and health

- Relying more on joint family systems (rather than small nuclear families), which are able to survive better amid climate change

- Building houses on safer ground, or raising them off the ground

- Building elevated tubal wells that keep the water from becoming salty in times of floods

- Cooperating with government efforts and various disaster alert mechanisms and groups

- Receiving new seeds to plant after floods.

In the aftermath of the big storms and floods, the women can now go out from their homes, in ways they could not before. Time and again, it was apparent that the women had been empowered and were organizing, for example, to plant trees and protect the forests: "For us, trees are life, so we will continue planting more: three trees for every child." They are also developing their own self-help and income-generating activities, which give the women new self-confidence, and inspire the men to do likewise. We heard of how it was the women who appealed to and successfully lobbied the government to get the fresh water they require.

We saw and heard how communities are adapting to life-giving forces from nature and pursuing mitigation, using appropriate technologies or systems:

- Adapting crops to make them saline tolerant

- Introducing prawn farming (although this can also pollute the water supply)

- Organizing local people to own and manage their own environment: rights-based access to water, land and forests

- Harvesting rainwater for irrigation and other purposes

- Constructing houses using stones instead of bricks, to keep temperatures lower

- Establishing village grain banks of a few bags of rice from which people can draw in their time of need and replace as they are able.

Participants from other parts of the world were able to compare and share their experiences and insights from living and working in other areas vulnerable to climate change. The Arctic area of Canada, where the Inuit peoples live, serves as a global "air conditioner" for tropical areas, but the ice there is melting dramatically. Many low-lying communities of Denmark are likely to be flooded in the future. In Tanzania, climate change is becoming as urgent an issue as are HIV and AIDS. It is connected with other patterns of injustice: older women (stigmatized as "witches"), were blamed and killed for the death of children who in fact were dying from waterborne diseases escalated by climate change. Now, with cleaner water available, the children are healthier, and the old women no longer suspect. In Bangladesh, a new kind of rice is being used to adapt to the later, crop-destroying monsoon season. Bushfires, which recently ravaged parts of Australia, provoke reflections on where was God in all of this? Not only is earth suffering all around the world, but also our interconnections are becoming more apparent.

Our call

Thus, to the rest of the Lutheran communion, we issue a call to confession and action. To be in communion with creation means to be in solidarity with those victimized by climate change, who inspire and motivate our commitment and actions to redress climate change.

For Christians, God and creation are distinct, yet on biblical bases, God could be imaged as lamenting or weeping on behalf of what is occurring under climate change. By disturbing the known balance of nature with the impact of climate change, have we disturbed the wisdom of God, the blueprint that orders and integrates the web of creation (see Job 28 and Prov 8)? By mutilating the face of the planet with climate change disasters, are we desecrating the sanctuary of the presence of God who fills the earth and continues to create in, with and under this cosmos (see Isa 6:3 and Ps 104)? By polluting the atmosphere with excessive greenhouse gases, are we polluting the very breath of God that animates and rejuvenates us and our planet home (see Gen 1:2, 2:7)? By causing adverse climate changes, we have wounded earth and created a condition that has increased the cries of creation, the pain of the poor and the sufferings of God.

To cut back significantly on the carbon emissions that contribute so severely to climate change, massive changes in policy, practices and lifestyle will need to be made, especially by those in the more affluent areas of the world. The villagers were well aware of how industrialization has contributed to climate change. But, what was striking were the strong awareness and intentional actions being taken by the people in these severely affected communities to change their practices in order to adapt to and to mitigate climate change disasters. Their empowerment to take responsibility for their lives, land and future is a strong witness to those who contribute disproportionately to the problem, and often feel unable to make changes. For example, large timber companies in the world continue to cut huge numbers of trees, but the women cooking in these villages are burning dried leaves and cow dung in order to avoid cutting down one more tree.

The purpose of this encounter was to "bear witness" to what is occurring in an especially vulnerable area of the world. The above witness is what we want to bring to the wider Lutheran communion, as well as ecumenical and civil society partners. Our recommendation is that the learnings and insights from this process might be expanded and further pursued prior to and at the Pre-Assemblies leading up to the 2010 LWF Assembly in Stuttgart. For example, for a brief period prior to the Pre-Assemblies, a small delegation might visit an area especially affected by climate change, and bring that witness to the Pre-Assembly, as part of a cumulative process moving toward the 2010 Assembly.

In addition, we support and encourage the various advocacy positions related to climate change that the LWF Council and member churches have already taken, encourage others to do likewise, and urge that there be a strategic presence and message of the LWF at the December 2009 UN Framework Convention on Climate Change in Copenhagen, Denmark. To coincide with that crucial meeting, we propose that a time be designated and promoted globally for ringing church (and other) bells in order to emphasize the urgency of redressing climate change.

> May God, the mother of the village well and the village women, help you draw water for life and laughter. May God, the father of the outcaste poor and deserted Dalits, meet you waiting in their streets and teach you hope. May Jesus, a son to malnourished mothers and a brother to unwanted daughters, teach you to be a midwife who brings new life from the risen one. May the Spirit, who seeks justice for earth oppressed by the ways of the past, lead you to open new eyes to see the path beyond evil to freedom. Amen.[3]

---

[3] Blessing at the conclusion of the Eucharist at Puri.

A Faith and Life-Changing Experience

Anupama Hial

The Puri encounter was an intense and challenging experience. Here, climate change has an existential impact on the lives of the poor and marginalized and presents a great challenge, especially to those of us coming from the state of Orissa.

The exposure to the realities faced by the community in Puri has been faith changing for me. As a pastor, serving in remote areas of Orissa, this experience has broadened my horizons and heightened my awareness of the importance of eco-theology. Throughout I was asking myself, What does it mean to be in the church and to be a pastor to people experiencing the life-changing and life-threatening impacts of climate change? I have been challenged to continue to learn from the people, and because of a new vision, to change my lifestyle. To be told by the communities that such simple actions as planting trees in the coastal areas can help the situation gave me new insights.

"The earth is the Lord's and all that is in it, the world, and those who live in it; for he has founded it on the seas, and established it on the rivers " (Ps 24:1–2). This verse helps us to understand that God alone, who created nature, has absolute control over creation, but at the same time, human beings are responsible for God's creation. This theological and ethical principle has been violated. Instead of preserving nature, we have exploited and plundered it. This has resulted in suffering—both for nature and for human beings. The worst affected in this process are the tribal and other marginalized groups. They suffer the most because of changing water levels, changing climate, salination, new diseases, etc.

Women are more deeply impacted by the effects of climate change. Being care providers and family sustainers, they need to ensure food security and the availability of clean drinking water, earn incomes to support the families, etc. In most cases, they walk long distances and spend considerable energy on fetching water and fuel.

The Christian response should be based the Spirit of Jesus, who came into the world to "make all things new." We need to confess how through greed and over-consumption we have degraded and exploited creation. The environment is not to be understood primarily in terms of resources (trees, air and water), but as an ecosystem with people's lives and needs at its center. The communities we visited have been deprived of their livelihood through ecological degradation. Their lives are no longer sustainable. Thus, the theological and biblical notion of "making all things new," does not merely imply restoring nature, but rebuilding the communities and relationships that have traditionally kept a balance between humans and the environment.

> Our God, the source of life, your name be glorified as you reveal yourself through the beauty of your creation.
> Let your will of protecting and preserving your earth community be continued.
> Give the poor the strength to own their forests, lands, seas and rivers to sustain them daily.
> Forgive us for breaking nature's cycle of life.
> Lead us not into temptations of greed, want and seeking to make a commodity of your creation.
> Deliver us from becoming evil agents destroying your creation.
> For yours is this cosmos which proclaims your reign and glory forever and ever.
> Amen.[1]

---

[1] Prayer inspired by the Lord's Prayer, prayed at the LWF encounter in India.

Discerning the Times: A Spirituality of Resistance and Alternatives

George Zachariah

Today, nature's most threatened creatures are not the whales or the giant pandas of China, but the poor of the world, condemned to die of hunger and disease before their time.[1]

You hypocrites! You know how to interpret the appearance of earth and sky, but why do you not know how to interpret the present time? (Lk 12:56).

Introduction

At its 2008 meeting in Arusha, Tanzania, the Council of the Lutheran World Federation (LWF) voted,

> to call upon member churches to engage in and deepen their theological and ethical reflection on the human contribution to climate change ... recognizing human beings as "co-creatures" with moral agency rather than claiming the prerogatives of creators, and seeking to learn from Indigenous practices and traditional wisdom for living sustainably as part of Creation.

This resolution not only demonstrates the communion's commitment to engage in transformative practices in the context of climate change, but also analyzes what is at stake.

Climate change is no longer an issue only for environmental activists and "green anarchists" from the Northern hemisphere. It is a global reality that communities—particularly those who are at the margins—experience in everyday life. Unprecedented variations in climate and global warming, frequent and intense devastation of land and communities due to severe droughts, floods and hurricanes, epidemics and new diseases, unmatched

---
[1] Leonardo Boff and Virgil Elizondo, *Ecology and Poverty* (London: SCM Press, 1995), p. x.

increases in the number of impoverished, displaced and exiled climate change refugees—these are the everyday experiences of communities all over the world. When survival requires might and money, conflicts become a daily ritual where the marginalized are sacrificed. Climate change has invaded us.

A few years ago, the American Museum of Natural History in New York organized an exhibition showing the different types of pollution. The last exhibit was a mirror labeled, "The Most Dangerous Animal on Earth." African American schoolchildren also came to see the exhibition, but their teacher had a tough time trying to explain the meaning of this particular exhibit as they stood in front of the mirror, blamed for the earth's pollution, while big corporations and governments are absolved.

Our prevailing discourse on climate change is like this apparently value free and apolitical mirror. In interpreting the appearance of earth and sky, we put up big mirrors in front of humankind and identify them as the root cause of the problem, by pronouncing that climate change is caused by the emission of anthropogenic greenhouse gases. Climate change thus becomes a discursive strategy to avoid interpreting the present time. This is an important statement and it should remain. Counter engagements with climate change discourses are needed from the perspectives of climate injustice. This article is an attempt in that direction, drawing on the experiences and spiritual resources of the subaltern communities and movements in India.

Analyzing climate change from the margins

In the process of carrying out the LWF survey on climate change,[2] we focused on the subaltern communities in India—the landless Dalits in Kerala, the Adivasis in Orissa and the fisherfolk in the coastal regions. They shared with us how they experience climate change in their lives. While the subaltern communities in India are yet to speak the language of climate change discourses, they are vocal in narrating how strongly climate change affects them in their daily lives. In recent years, climate patterns have changed drastically. As subsistence communities, dependent on *jal, jungle* and *jamin* (water, forest and land) for their daily sustenance, their future existence is under threat. The unpredictability of monsoons and the unusual distribution of rainfall have upset their agriculture and farming practices. Changes in the monsoon season and the depletion and

---

[2] See p. 11 in this publication.

poisoning of underground water resources not only affect the availability of drinking water, but also the life cycle of plant life and the ecosystem. This has meant that the communities have lost food sovereignty. The frequent outbreak of vector borne diseases such as Chikungunya and dengue fever further torment their lives. As a result, the subaltern communities have been uprooted from their organic habitats: the forests, the land and the coastal regions and have become refugees in their homeland. As one old fisherman observed, climate change has even disrupted their sense of time. Alienated from time and space, today's climate refugees live in the slums of big cities such as Mumbai, Kolkata and Chennai, the very cities that face deluge due to climate change.

Climate justice as economic justice

A deeper engagement with the symptoms of climate change described by the subaltern communities invites us to analyze the issue in a different way. Such an attempt stems from the gospel imperative to discern and interpret the present time. As Michael Northcott rightly puts it, climate change is "the earth's judgment on the global market empire."[3] Put differently, climate change is the consequence of the colonization of our lifeworlds by imperial, capitalist and neoliberal projects and their systematic interventions. The impact of the "judgment of the earth" is disproportionately suffered by the communities at the margins, due to the sinful social relations of casteism, patriarchy and economic injustice, which are perpetuated with religious legitimation by those who dominate. Such discernments provide us with the impetus to initiate new cognitive, theological and political praxis to resist and to create alternatives.

In 2001, the Third Annual Report of the Intergovernmental Panel on Climate Change (IPCC) categorically stated that,

> the impacts of climate change will fall disproportionately upon developing countries and the poor persons within all countries, and thereby exacerbate inequities in health status and access to adequate food, clean water, and other resources.[4]

---

[3] Michael Northcott, *A Moral Climate: The Ethics of Global Warming* (Maryknoll: Orbis Books, 2007), p. 7.

[4] At **www.ipcc.ch/ipccreports/tar/vol4/english/009.htm**, accessed June 2009.

This forecast has already been realized in the life stories of the subaltern communities in India. According to official statistics, due to the commodification of land, water and forest by neocolonial capitalist forces, the number of internally displaced people in India is over fifty million. They are the climate and environmental refugees; the victims of climate injustice. Their ethnic otherness reveals the class, caste and gender bias of climate injustice.

The subaltern analysis of climate change in India is significant in the context of the global climate change discourse where "developing" countries such as India try to absolve their sins of emission by using the rhetoric of social justice and the "right to develop." When we measure development in terms of GDP, we require more mega dams that denude rain forests and displace Dalits from their livelihood, new coal mines that stretch over the landscape of the Adivasis, four lane expressways and new airports that convert paddy fields into runways and roads that bring in foreign investment. As a nation we courageously affirm that nobody can deny us our right to develop, and hence carbon emissions are morally justifiable. This spirit is the rationale behind Prime Minister Manmohan Singh's statement that "India's per capita emission level will never be higher than the Western countries."[5] As Arundhati Roy opines, "India's poorest people are subsidizing the lifestyles of her richest."[6] This situation compels us to raise fundamental questions such as, Who are the "we" in this narrative? Who owns this India that our leaders represent? Whose "development" it is anyway?

As the online journal *Mausam* categorically affirms,

> It is time to say loudly that the crisis is not really about climate. It is not about rising sea levels and the melting arctic, dead seals and polar bears facing extinction. It is about us, our lives, and the planet—and the way the powerful and rich of the Earth have dominated and kept destroying them for centuries, to accumulate private wealth.[7]

---

[5] At **http://southasia.oneworld.net/Article/equity-key-to-global-warming-crisis**, accessed June 2009.

[6] Arundhati Roy, *The Greater Common Good* (Bombay: India Book Distributors, 1999), p. 11.

[7] *Mausam ... Talking Climate in Public Space*, vol. 1, issue 1 (July–September 2008), p. 1, at **www.thecornerhouse.org.uk/pdf/document/Mausam_July-Sept2008.pdf**, accessed June 2009.

A subaltern perspective further exposes the prevailing practice of portraying climate refugees as victims, in order deliberately to exclude them from the cognitive and political processes to address and solve the crisis. This categorically calls for a fresh look at the causes of global warming and climate change from the standpoint of the climate refugees.

In his speech releasing the National Action Plan on Climate Change on 30 June 2008, Prime Minister Singh drew on the panentheistic tradition of Indian civilization to legitimize his market driven and growth oriented neoliberal action plan on climate change:

> India has a civilizational legacy which treats Nature as a source of nurture and not as a dark force to be conquered and harnessed to human endeavor. There is a high value placed in our culture to the concept of living in harmony with Nature, recognizing the delicate threads of common destiny that hold our universe together. The time has come for us to draw deep from this tradition and launch India and its billion people on a path of ecologically sustainable development. Our people have a right to economic and social development and to discard the ignominy of widespread poverty. For this we need rapid economic growth. But I also believe that ecologically sustainable development need not be in contradiction to achieving our growth objectives.[8]

As we engage with climate change, it is essential that we critically analyze this affirmation of faith in the redemptive potential of neoliberal capitalism. Prevailing discourses on climate change not only refuse to recognize the integral relationship between capitalism and climate change, but also propose the neoliberal economic model as a panacea, namely ecologically sustainable rapid economic growth. Let us look at some concrete examples of the growth oriented development projects in India and their impact on climate change to see whether climate change can be arrested within the prevailing trajectory of capitalist economic development.

Automobiles contribute a major part of the greenhouse gas emissions in India. Chennai is considered India's Detroit. A forty-five-kilometer corridor on the outskirts of the city has become one of the largest automobile centers in the world. Hyundai, Nissan, Renault and Ford have already started their manufacturing units here. It is estimated that by 2012 Chen-

---

[8] At **www.pmindia.nic.in/speech/content.asp?id=690**, accessed June 2009.

nai alone will produce 1.28 million cars, 350,000 commercial vehicles and a large number of tractors and earth moving equipment every year.[9]

The state of West Bengal was the site of the historic struggle by local communities against being forcefully evicted from their agricultural land by the government in order to construct a plant to produce TATA Nano, the world's cheapest car, priced at USD 2,500. In order to invite capital investment in the state, the government offered an estimated subsidy of Rs 8,500 million for an investment of Rs 10,000 million. The communities succeeded in closing down the plant. However, the government of Gujarat offered the company a better deal, and the company is going ahead with the project in Gujarat. This project is envisioned as an example of India's growth and progress because it would revolutionize the automobile sector and the lifestyle of millions of middle-class Indians. However, the rhetoric of growth and progress ignores the truth that such projects not only significantly increase greenhouse gas emissions, but also uproot communities from their land and livelihood.

In India's tryst with destiny, it was Prime Minister Jawaharlal Nehru who paved the way for progress through industrialization and mega development projects. Nehru considered dams to be the modern temples of the nation. By calling economic models and development projects that colonize the lifeworld "temples," he initiated a new religiosity that considers capitalist development as a fetish. As we look back after six decades of independence, mega dams stand out as the greatest culprit for the ecological crisis in India. While in climate change discourses, hydroelectric energy is considered clean, a recent study concluded that nineteen percent of India's global warming emissions are from the methane emissions of the mega dams. The study estimates that total methane emissions from India's large dams could be 33.5 million tons (MT) per annum, including emissions from reservoirs (1.1 MT), spillways (13.2 MT) and turbines of hydropower dams (19.2 MT). Climate policy makers, including the IPCC, have largely overlooked the importance of dam generated methane in their climate change discourses.[10]

---

[9] At **http://businesstoday.digitaltoday.in/index.php?option=com_content&task=view&id=6848&issueid=39**, accessed June 2009.

[10] For over a decade, large dams have been known to be emitters of greenhouse gases such as methane, carbon dioxide and nitrous oxide. The "fuel" for these gases is the rotting of the vegetation and soils flooded by reservoirs, and of the organic matter (plants, plankton, algae, etc.) that flows into. Methane is produced at the bottom of the reservoirs in the anaerobic conditions prevailing there, over the lifespan of the reservoirs. The gases are released at the reservoir surface, at turbines (of hydropower projects) and spillways, and downstream of the dam. For more information, see South Asia Network on Dams, Rivers & People, New Delhi, at **www.sandrp.in**, accessed June 2009.

A counter engagement with climate change, informed by the narratives of climate refugees, challenges us to see the intrinsic relationship between neoliberal interventions in nature and climate change.

Prevailing policies and strategies to address the crisis posed by global warming seem to be rooted in the faith in neoliberal capitalism. Quests to substitute fossil fuel with biofuel and other energy sources such as nuclear power, carbon trading, carbon taxation, Clean Development Mechanism (CDM) and the like are embedded in the logic of the prevailing trajectory of economic growth as development. Climate change discourses appear to be geared toward finding solutions within capitalism.

We need to have the discernment of Audre Lorde boldly to proclaim that the "master's tools will never dismantle the master's house."[11] When corporate interests dominate climate change discourses the alternatives that emerge from the dominant discourses can only be a continuation of profit oriented interventions with nature, packaged in new attires with prefixes such as "bio" or "green" added. We have yet to discern them as what Naomi Klein calls "disaster capitalism."[12]

On 12 September 2008, the government of India introduced a new national biofuel policy, which aims by 2017 to meet twenty percent of India's demand for diesel with fuel derived from plants rather than fossils. Fourteen million hectares of land are to be set aside for producing biofuel. Jatropha is the main shrub that is planted in India in collaboration with D1 Oils and BP. Jatropha contains a toxic protein and is considered a poisonous plant. However, India accounts for about two-thirds of the world's jatropha plantations. In the context of the recent food shortage, the Finance Minister of India described the conversion of food crops into biofuel as "a crime against humanity."[13] But we continue to convert our agricultural land into biofuel plantations in the name of ecologically sustainable rapid economic growth. On 7 November 2006, when the former President of India, A. P. J. Abdul Kalam, came to Chhattisgarh to inaugurate the jatropha plantation, he was welcomed with the slogan: Welcome to Chhattisgarh, the land of jatropha. Chhattisgarh is predominantly an agrarian society of landless Dalits and Adivasis whose livelihood depends on rice cultivation. In an open letter to the former President, subaltern movements in the state affirmed that

---

[11] Andre Lorde, *Sister Outsider: Essays and Speeches* (Berkeley: The Crossing Press, 1984), p. 110.

[12] Naomi Klein, *The Shock Doctrine. The Rise of Disaster Capitalism* (New York: Picador, 2007).

[13] "The slow ripening of India's bio-fuel industry," in *The Economist* (20 September 2008).

any reference made to Chhattisgarh as the "land of jatropha" undermines the significance of "rice" as the foundation of people's economy, cultural identity, and dignity and is an insult and open attack on their rights to life and livelihoods.[14]

The neoliberal solution of biofuel systematically deprives communities of their food sovereignty. Moreover, as the subaltern movements rightly identify, the shift from rice to jatropha is an attempt to humiliate the Dalits and Adivasis by destroying their cultural identity, cognitive ability and communal practices.

The planning commission proposes large-scale jatropha cultivation starting with 400,000 hectares of land in the first phase, which will eventually be extended to 13.4 million hectares. The areas identified for this ambitious mission are the wastelands along with cultivable fallow and barren lands. The planners failed to recognize that the so-called wastelands or fallow lands are in fact village commons or common property, resources that are intrinsically connected with the livelihood, cultural identity, indigenous religious beliefs and practices and food sovereignty of the subaltern communities. By incorporating the village commons into the logic and rules of neoliberal capitalism, the biofuel industry disables the moral agency of the subaltern communities through pauperization and ethnocide. Solutions to climate change borne out of the logic of capitalism can never save the earth or its children from the bondage to decay.

Clean Development Mechanism (CDM), a project to assist the global South to achieve sustainable development, is yet another attempt to find solutions for climate change within the neoliberal paradigm. It provides opportunities for Northern corporations to buy carbon credits and to continue to pollute. They are the contemporary manifestations of indulgences to absolve the carbon sins of the corporations. Under the carbon sink projects, corporations invest in tree plantations in Southern forests. But they plant millions of trees of the same species—industrial agricultural crops—and breed them for rapid growth and high yield. The communities in Brazil are resisting such monoculture tree plantations with the slogan, *diga não ao Deserto Verde* (Say no to green deserts). In the context of the food crisis, the conversion of agricultural land into biofuel plantations shows a shift in government policy from feeding

---

[14] At **www.biofuelwatch.org.uk/docs/mausam_colonizingthecommons_itsjatrophanow.pdf**, accessed June 2008.

communities to feeding cars. Carbon sink projects do not respect the customary rights of the indigenous communities to their land and livelihood. All these contribute to new exoduses of climate refugees, and take away the food sovereignty of the communities. A counter engagement with climate change, therefore, invites us to a new journey, which is a fundamental departure from the prevailing global order, and a radical transformation of our social, economic, political and ecological relations at local, national and global levels.

Closer scrutiny reveals the prevailing climate change discourses as conscious attempts to absolve the sins of the neoliberal capitalist plunder. By initiating programs and projects to mitigate and adapt climate change, ignoring the very fact that these "alternatives" are embedded in the colonizing paradigm and logic of economic growth, the culture of death and destruction is perpetuated. Such "alternatives" promise affluent societies the opportunity to continue to enjoy high standards of living and consumption patterns if they shift to non-fossil fuel energy sources. The implication is that technical solutions to climate change within the logic of capitalism will not only make mitigation and adaptation painless but also profitable. Pay and pollute models of indulgences and the search for alternative sources of energy lack the boldness to say "no" to affluent consumption patterns and lifestyles. Instead, what is needed is a radical shift from the prevailing dominant analyses to seeing climate change as a problem of injustice.

Climate justice as gender justice

Mainstream climate change campaigns have been sensitive to bringing the gender dimension into their discourses. However, gender sensitivity does not always stem from a political commitment to gender justice but from a diplomatic attempt to be politically correct. Subaltern communities contest and reject this strategy of "add and stir" and affirm the interconnectedness between climate justice and gender justice. To put it differently, climate justice is not possible without gender justice.

Conservative statistics reveal that about seventy percent of the world's poor, who are especially vulnerable to environmental disasters, are women; they are more likely to suffer due to climate change. Eighty-five percent of those who die in climate-induced natural disasters are women. Seventy-five percent of environmental refugees are women.

Moreover, women are also susceptible to being the unseen victims of resource wars and conflicts resulting from climate change.

What are the reasons for this gender bias? Prevailing social inequalities legitimize and perpetuate the disabling of women's moral agency through lacking educational and employment opportunities, patriarchal property rights, a lack of access to information, physical and sexual abuse and violence, etc. These make women more vulnerable to the impacts of climate change. Sectors affected most by climate change are sectors traditionally associated with women, such as rice paddy cultivation, cotton and tea plantations, fishing (women in India are actively involved in the fisheries sector), etc. Thus, women are confronted by situations in which their ability to adapt is low due to gender constructions and prevailing gender relations, and are asked to bear the burden of adaptation disproportionately.

Adaptation is a means of reducing vulnerabilities caused by global warming and other ecological disasters, so that communities that disproportionately bear the brunt of the problem, can be protected by empowering their moral agency. However, adaptation to climate change is dependent on one's wealth, technological power, access to information and social standing. Gender plays a major role here. The integral connection between gender justice and climate justice must be recognized and gender factors incorporated when assessing climate change.

In traditional approaches to "women in development," women are perceived as victims, and measures are being taken through which women become the objects of development. Women themselves challenge this approach by exposing the top-down power dynamic of this model. Further, in this approach women are represented as a homogenous entity, without recognizing the multiple forms of oppression that women undergo. The experience of a particular class of women is universalized, i.e., as the experience of all women. A subaltern perspective on climate change contests this approach and proposes that women be recognized as subjects with moral agency, cognitive power and the political will to address and transform climate injustice.

Climate justice as justice for Dalits and Adivasis

From the perspective of subaltern communities, climate change exposes the connection between climate injustice and the casteist social relations in India. As we have already seen, the skin color of the environmental refugees reveals the caste bias of the climate-induced and other environmental disasters in India. The caste system, being the original sin of this land, is deeply entrenched in India's social fabric, and hence the victims and refugees of climate change and other environmental disasters are not an abstract category of people without faces and names; rather, they are predominantly Dalits and Adivasis. This discernment is totally absent in the official, dominant climate change discourses in India. This calls for a greater engagement of the Dalit communities and movements in assessing the problem of climate change from the vantage point of their experiences.

In attempting to develop a Dalit/Adivasi perspective on climate change, the recent study entitled, *A Climate of Change: African Americans, Global Warming, and a Just Climate Policy for the U.S.*,[15] is a helpful resource. This study views climate change not only as an environmental issue, but also a matter of justice and human rights that "dangerously intersects race and class."[16] This enables us to understand how climate change contributes to perpetuating prevailing social inequalities.

A Dalit/Adivasi perspective on climate change proposes a shift in how we see the problem. Climate change is more than variations in temperature; it is an environmental crisis that disproportionately affects Dalits and Adivasis. Yet, these climate refugees are not the culprits. While their carbon footprint is insignificant when compared with the dominant caste and class in India, when it comes to the impact of climate-induced disasters, they are the ones who are destined to suffer the most. Their social exclusion and economic dependency are reinforced through climate change. They are displaced from their traditional land-centered occupations, and find themselves in the dehumanizing reality of being exiled and selling their lives to survive.

Engaging climate change from the experiences of the Dalits and Adivasis affirms the agency of primal communities who from time immemorial have lived in harmony with land, jungle and water. Attempts to save forests, land and water from the conquest of climate change will become meaningful only when linked with the struggles of people such as the Dalits and Adivasis for identity, equality, dignity and freedom.

---

[15] J. Andrew Hoerner and Nia Robinson, *A Climate of Change: African Americans, Global Warming, and a Just Climate Policy for the U.S.*, at **www.ejcc.org/climateofchange.pdf**, accessed June 2009.

[16] *Ibid.*

When their moral agency is empowered, they become effective agents for climate and social justice.

An ethical imperative for the church

Climate change is a global reality which can no longer be ignored in the church's social witness. However, climate change raises several profound questions that demand theological and biblical rethinking. According to Karen L. Bloomquist,

> Climate change may literally be melting icebergs but it also exposes metaphorical icebergs of how God, human beings and the rest of creation have been conceptualized in ways that contribute to the injustices that have only increased under the currently reigning realities of climate change.[17]

Where is God in the midst of climate-induced tragedies? Has God left us alone and totally disappeared from the scene? Is climate change God's punishment for our transgressions? How do we explain the biblical texts that attribute nature's fury to God? How can we chastise human beings for being anthropocentric when we have been given the responsibility to subdue the earth? Climate change invites us to wrestle with these questions in order to equip ourselves to respond to the crisis caused by climate change.

Martin Luther understood creation as God's abode. He did not experience God as an impassible deity detached and removed from the world of creation, but rather as being in, with and under all that is creaturely. The mystery of creation is God's indwelling within it. Luther helps us to reimagine God as the indwelling spirit within the created world, despite the tragic experiences of climate injustice. Luther's theology of the cross is also insightful as we confront experiences of utter God forsakenness in the context of climate-induced disasters. To rephrase Luther, God is hidden in the pain and agony of the climate victims and climate refugees. The crucified God, whom we meet on the cross, is present in our life stories, even as we go through similar experiences due to climate change and other environmental crises. God has not deserted us; God is a fellow

---

[17] Karen L. Bloomquist, "Setting the Scene for the Consultation," a background paper for the LWF Consultation on Climate Change held in Geneva in October, 2008.

climate change refugee in exile, keeping our hope alive in the midst of hopelessness. Incarnation is God's embracing of the material world; the Word became flesh and dwelt among us. Redemption from John of Patmos's perspective is not a rescue operation of the saved souls, but rather the vision of a redeemed earth. The church is called to witness to the inbreaking of this vision in the here and now.

Climate change: a call for social metanoia

"Climate Change—Vulnerability, Lament and Promise" was the theme of the 2008 LWF Sunday. Barbara Rossing, preaching on that occasion at the Lutheran School of Theology at Chicago, reinterpreted the story of Jonah and Nineveh from the perspective of climate change. The great imperial city of Nineveh was asked to repent and mend its ways. God had given the city forty days to do so. God did not want Nineveh to be destroyed, but wanted Nineveh to turn and to repent. God wanted to save Nineveh. Rossing identified two insights in the story of Jonah that are of relevance to climate change. First, God is not a God who intends to punish us for our inequities. Rather, God is a God who loves God's creation. God does not want creation to be destroyed. Realizing that the great imperial capital city could turn back to God and mend its ways assures us that we can today make climate change history through our acts of repentance and by mending our ways.

The world in which God dwells is the site of grace and healing. Nineveh is a paradigm for all of us. The unbelievable good news that we find in the story of Nineveh should inspire us to journey from a politics of cynicism and despair to a politics of hope—the audacity of hope (as President Barack Obama puts it). But it demands from us a costly commitment to undergo a social *metanoia*.

Rediscovering the communities' spiritual resources

Analyzing climate change from the standpoint of the subaltern communities exposes climate change as a crisis of meaning, purpose and vocation. It is a deeply spiritual problem, integrally connected with our relationship to the divine, the human community and the wider community of creation. Stated differently, climate change is the consequence of a faith which absolutizes the neoliberal market mechanisms as realized eschatology, propagates an anthropology that understands fulfillment of self-interest as human flourishing, approaches nature as a bounty given to humankind to grab and to control, and believes in a god who sanctions the sacrifice of human and other lives for the prosperity and well-being of the chosen ones. This is the spiritual crisis that we face in our context. We must turn to the subaltern communities for new resources that can help us in our search for spiritualities that inspire and empower us to decolonize our minds, our faiths, our communities and our planet.

Spirituality, for Konrad Raiser, "refers to that source of energy which generates and regenerates a sense of purpose and recognition of values in social life and thus nurtures the social fabric."[18] The Jesuit theologian, Samuel Rayan, interprets spirituality as "openness and response-ability to others."[19] Decolonizing our faiths and our times requires spiritualities that provide us with the creative energy to reimagine our sense of purpose as created co-creators, so as to be open with "response-ability" to the subordinated others, the subalterns. This includes the spiritual discernment to perceive our earth as a subaltern earth. The subaltern communities in India provide us with diverse spiritual resources that can challenge and inspire us in our search for meaning and purpose.

The Ashur myth from the Munda tribe of Central India is of great significance as we reflect on spirituality in the context of climate change. The Ashurs were early industrialists. They continuously used furnaces for smelting iron, which led to the warming up of earth, the home of all creatures. When the warming of the earth became unbearable for the animals, the insects, the trees and all other creatures, they approached Singboga, the sun god, to ask for his intervention and to persuade the Ashurs to stop smelting iron. God heard their cry and sent birds as messengers to the Ashurs. The birds told them that the god wanted the Ashurs to restrict the smelting of iron either to the day or to the night. That was god's compromise in order to keep the industry going without affecting the creatures. But the greedy Ashurs did not listen

---

[18] At www.oikoumene.org/en/resources/documents/wcc-programmes/public-witness-addressing-power-affirming-peace/poverty-wealth-and-ecology/neoliberal-paradigm/12-09-03-spirituality-of-resistance.html, accessed June 2009.

[19] Samuel Rayan, "The Search for an Asian Spirituality of Liberation," in Virginia Fabella *et al.* (eds), *Asian Christian Spirituality. Reclaiming Traditions* (Maryknoll, NY: Orbis Books, 1992), pp. 25f.

to god's messengers. Finally, the god himself came down to earth as a boy. He came to the Ashurs and requested them to stop smelting iron either during the day or the night. They got angry, caught him and threw him into the furnace. To their surprise, the boy rose from the furnace with gold and diamonds on his body and in his hands. Since the greedy Ashurs wanted to grab all the precious metal from the furnace, they all jumped into the furnace and asked their wives to bellow faster. Quickly the wives realized that their husbands were being burnt to ashes. They wanted to take revenge on the boy, who by that time was being lifted up into the sky. The wives tried to catch hold of his hands and feet, but all fell down on the burning earth.[20]

The Ashur myth envisions the earth as the household of all creatures. Human beings are not given any special privileges in this assembly of creation. Rather, they all strive together for the welfare of all. The vision of God in the Ashur myth reveals a God who hears the cry of all, even non-human beings. The messengers of God—the prophets—are the birds. In response to the cry of creation, God becomes incarnate in their midst as a young child—the subaltern one—to redeem the earth and its children from global warming. In this redemptive incarnation and praxis, God shares the pain and agony of God's beloved creation and is burned alive by the forces of death. But God transcends the shackles of death and emerges as the reigning symbol of eternal hope and life by defeating and destroying the agents of death. The forces of greed shackle the creativity of science and technology and make it subservient to the interests of mammon. But the divine "eco-sophy" has the creative potential to transform the very furnace of global warming into a site for a new politics and spirituality of resistance and victory over death and destruction. This resistance against the powerful is led by a young child with the support and solidarity of the assembly of creatures. The image of transforming the furnace of death into a resource pool for the prosperity and well-being of the whole inhabited earth is a powerful vision inspiring us to believe in the possibility of a world redeemed from climate change.

---

[20] Wati Longchar, *An Emerging Asian Theology: Tribal Theology: Issues, Method and Perspective* (Jorhat: Eastern Theological College, 2000), pp. 70–71.

The spirituality of resistance and alternatives

For those who struggle for survival, spirituality is the spirit-filled energy that enables people to believe in alternatives, as they continue to resist the idols of death. This belief sustains them in their struggles. As Leonardo Boff rightly puts it, it is "the advent of divine redemption mediated through historical-social liberations, the moment when the utopia of integral liberation is anticipated under fragile signs, symbols and rites."[21] It is an option for life, a confrontation with the logic of death and the celebration of life. This is the testimony of subaltern communities and their social movements.

Subaltern communities and movements are the seedbeds of regeneration and recovery. As geographies of difference, they see reality differently and search for alternative ways of engaging with it. They are cognitive communities that are involved not only in critiquing prevailing models of technology, but also in developing their knowledge systems and practices as alternatives. The commonality in the struggles of the subaltern movements is their radical commitment to reclaim the commons. The struggle against the TATA Nano plant in West Bengal, the historical struggle in the Narmada valley against mega dams, people's struggle against the Coca Cola plant in Plachimada, the Dalits' Chengara struggle for land and livelihood, the farmers' struggle against genetically modified crops and the struggle of the fisherfolk in coastal regions–all these testify to people's resilience against the colonization of their commons. We can discern that climate crisis is due to the expropriation of the commons by capitalism, which displaces and disempowers people of color, women and indigenous communities.

What is the mission of the church in this context? God is at work in our midst, inviting people to turn and to repent. There are glimpses of this social *metanoia* in our villages and small towns. Our responsibility at this historic moment is not to Christianize the movement of social *metanoia*, but to enable and inspire our faith communities to join this movement in order to enhance life. This involves reengaging with the Bible and our faith tradition, and to reimagine liturgy and *kerygma* so that our faith communities will discern climate change as a wake-up call to repent. Energy audits, introspective lifestyle practices and reducing carbon footprints are effective means to lead us to the experience of

---

[21] Leonard Boff, *Cry of the Earth, Cry of the Poor* (Maryknoll: Orbis Books, 1997), p. 110.

social *metanoia*. But the primary responsibility of the church in India and elsewhere is to instill in our community the belief that if Nineveh can change, we too can change.

As the 2009 declaration of the Climate Justice Assembly in Belem, Brazil, states:

> For centuries, productivism and industrial capitalism have been destroying our cultures, exploiting our labor and poisoning our environment. Now, with the climate crisis, the Earth is saying "enough." Once again, the people who created the problem are telling us that they also have the solutions: carbon trading, so-called "clean coal," more nuclear power, agro-fuels, even a "green new deal." But these are not real solutions; they are neo-liberal illusions. It is time to move beyond these illusions. Real solutions to the climate crisis are being built by those who have always protected the Earth and by those who fight every day to defend their environment and living conditions. We need to globalize these solutions.
>
> For us, the struggles for climate justice and social justice are one and the same. It is the struggle for territories, land, forests and water, for agrarian and urban reform, food and energy sovereignty, for women's and worker's rights. It is the fight for equality and justice for Indigenous Peoples, for peoples of the global South, for the redistribution of wealth and for the recognition of the historical ecological debt owed by the North.
>
> Against the disembodied, market-driven interests of the global élite and the dominant development model based on never-ending growth and consumption, the climate justice movement will reclaim the commons, and put social and economic realities at the heart of our struggle against climate change.[22]

---

[22] At **www.climateimc.org/en/press-releases/2009/02/04/climate-justice-assembly-declaration**, accessed June 2009.

Caminhada:
A Pilgrimage with Land, Water and the Bible

Elaine Gleci Neuenfeldt

In Latin America, the concept of *caminhada* is very popular. It can have several purposes. It is a kind of pilgrimage, and as such can have religious connotations. As a political action, it can be used to initiate a social movement and to raise awareness regarding health or ecological issues (e.g., diabetes or in defense of a river). *Caminhada* is also associated with certain movements that protest against the realities of poverty and oppose the injustices that perpetuate it. One example is the *Romaria da Terra* (pilgrimage of the land), organized by the Comissão Pastoral da Terra (CPT), the Pastoral Land Commission, whose specific foci are: the "social function of the land;" "youth and the access to land;" "urgency of agrarian reform;" "access to and use of water." The *Romaria da Terra* is religiously inspired, but involves social and political engagement.

In this essay, I use the concept of *caminhada* to reflect on the effects of climate change. At each "stop," I shall analyze the context and reflect upon the implications for theology and spirituality. This theological "walking" can provide different tools to discuss sacred symbols, space or geography. This is aligned with a popular reading of the Bible, a methodology widely used in Latin America, that permeates these reflections.

First movement: access to and use of land

Every year, on Independence Day (September 7), Brazilians take to the streets to protest against social and economic injustices resulting from the unjust political and economic (neoliberal) system. This popular, public demonstration, organized by social movements, is called *O Grito da Terra*, the cry of the earth, and includes all those affected by social and economic injustice in rural as well as urban areas. It is also referred to as "The cry of the excluded." It is as if the earth were yelling, crying and screaming because of the brutal realities of poverty, marginalization and injustice and their effects on human life and the environment.

One of the reasons for this annual demonstration is how the productive land available for agriculture is distributed and used. Brazil is the largest country in South America. Its territory spans over 8,547,403 km², but the productive land is in the hands of only a few farmers. The distribution of land is unequal and therefore unfair. Approximately forty-seven percent of the land is owned by just one percent of the population, making the country's land distribution one of the most inequitable in the world. Of equal concern are the use, misuse and abuse of the land. For instance, the sugar-alcohol and paper-cellulose monopolies devour thousands of miles of productive land, destroy native forest and seeds, create a green desert and pollute rivers.

Furthermore, huge sums are invested in the production of energy based on the industrial processing of vegetables such as maize, soya bean and sugar cane. This kind of investment in the production of biofuel jeopardizes food production and, moreover, leads to the misconception that agrofuel (or so-called biofuel) can help prevent climate change. The production of biofuel leads to monoculture, the results of which are concretely experienced in the lives of the rural population: hunger, poverty, deforestation, the drying up or contamination of water sources, small farmers forced to leave the land, rural violence, increasing numbers of marginalized and poor in cities (*favelas*) and the creation of more *latifundios*, large landholdings, by a few owners.[1]

The women's movement in Brazil is one of the strongest, organized voices on the topic of access to and use of the land and the dramatic consequences of climate change. On 8 March 2002, approximately 2,000 women from La Via Campesina occupied the plantation of Aracruz Celulose, Rio Grande do Sul, in order to denounce the social and environmental impacts of the growing green desert created by the enormous plantations of eucalyptus, acacia and pines that cover thousands of hectares in Brazil and the rest of Latin America. Since 2002, on International Women's Day, women from various social movements remember and denounce the connections between political and economic power and the disastrous consequences of the way in which the government is beholden to the private sector.

Gender inequalities and climate change are connected. "The women acted in defense of life, rural development based on the agricultural

---

[1] More information, at **www.cpt.org.br/index.php**, accessed June 2009.

family structure, preserving biodiversity and building up of food sovereignty."[2] "Women are important actors of change and holders of significant knowledge and skills related to mitigation, adaptation, and the reduction of risks in the face of climate change, making them crucial agents in this area.[3]

Second movement: walking with the Bible

By reading the Bible and understanding how it is related to daily life, the second movement turns to the Bible for light and inspiration. For many Latin American communities, the Bible is a sacred text. The study of the Bible from popular, community and ecumenical perspectives has provided the poor and excluded with a chance to regain their self-esteem and to become full citizens. This way of reading biblical text can illuminate new perspectives for a faith-based approach to climate change. The intent is to draw connections between some biblical testimonies and the experience of social movements, especially women's movements, as a starting point for theological reflection.

These experiences then are interwoven with theology and spirituality in times of climate change. What does this mean theologically and pastorally? What kinds of theological questions need to be asked? How can we come to grips with the complexity of this context? How does the need for food sovereignty and access to land challenge our theology?

We begin with the prophetic context in Isaiah, and search for biblical metaphors that can empower women and men. This is an exercise in reading biblical texts in a time of climate change, seeking changes in our theology and biblical approach that are motivated by the challenges and possibilities that arise from this reality.

The road of the land

With biblical inspiration or/and conspiration,[4] crying and struggling in the midst of climate change suggests turning to the prophecy in Isaiah

---

[2] More information, at **www.mmcbrasil.com.br**, accessed June 2009.

[3] *Training Manual on Gender and Climate Change*, at **www.reliefweb.int/rw/lib.nsf/db-900sid/ASAZ-7SNCA9/$file/UNDP_Mar2009.pdf?openelement**.

[4] Co-inspiration, inspiring/breathing together.

5:8–10. The prophet Isaiah denounced the injustice against the land, but also the rising political and economic interference from the Assyrian empire. From/with the land, the prophet criticizes the notion of prosperity and peace that was upheld through the official temple religion, and had the effect of placating and pacifying the poor and oppressed.

The prophecy in this text is directed against the wealthy landowners, who are treating the Promised Land as their property, accumulating it, selling it and not respecting the law. In Leviticus 25:23, it is clearly stated, "The land shall not be sold in perpetuity, for the land is mine; with me you are but aliens and tenants." The Jubilee tradition expressed in 25:10 ("you shall return, every one of you, to your property...") is in opposition to the land being owned by only a few and used unproductively. The consequence of accumulating land is less production and less food.[5] Here, the prophetic word recalls the connection of the land with God. The way in which land is treated is not only a socioeconomic issue but also a spiritual concern. How people deal with land and production is connected with how they understand and live out the Word of God, which is based on justice and equality. The prophecy calls for justice, land, food and a home for all, and is thus contrary to the understanding that God provides prosperity only for the chosen few.

Prosperity theology, which is widespread in our times, is based on the global market: religion, spirituality and theology can be consumed as if they were goods. Consumerism is the way to fulfilling desires and searching for meaning. Religion shaped or framed by this ideology is empty of hope for the poor. It implies that the poor are poor because they are not blessed by God, that catastrophes and diseases are a punishment from God and that God punishes and penalizes those who cannot enter the world of those chosen to be winners. It is a theology without grace, marked by tragedy on the one side and glory on the other. God will bless all those who are able to go ahead in this way. This theology is in synchrony with the market. It fits with a neoliberal ideology and will not advocate for justice and integrity of life.[6]

Isaiah's prophecy helps us critique this theology and to look for other possibilities in terms of a spirituality rooted in the daily life of the poor. There is a critical principle in the Bible: God denounces injustice. In

---

[5] J. Severino Croatto, *Isaías. A palavra profética e sua releitura hermenêutica*, vol. I:1–39 (São Paulo: Vozes, Sinodal, Metodista, 1989), pp. 52–55.

[6] See Oneide Bobosin, *Correntes Religiosas e Globalização* (São Leopoldo: CEBI, PPL, IEPG, 2002).

Isaiah 5:8, injustice resulting from the concentration of land and property in the hands of a few is denounced "Ah, you who join house to house, who add field to field, until there is room for no one but you, and you are left to live alone in the midst of the land!" For instance, *latifundios* can result in a decline in the production of food.

God sees those who cry out because of their suffering, and is not indifferent toward the suffering and cries of the poor and oppressed. Exodus 3 assumes a spirituality that is inserted into and committed to the transformation of unjust situations. In various passages in the prophetic books there are references to the experience of the exodus (the road out) as a road of liberation.

It is clear that the discussion on climate change is concretely connected with access to and use of land. Climate, environment and ecology related issues are linked to the question of access to and use of land. It is a social, political and economic issue and provides a key to engaging with the biblical text.

The road of the water

Another important aspect of climate change is the use of water, which is closely related to the production of food. In the Bible, the theme of water is expounded upon in the prophecy of Second Isaiah. Isaiah 43:16–21 promises the unlikely: "I will make a way in the wilderness and rivers in the desert ... for I give water in the wilderness ..." (Isa 43:19–21).

The theme of the road reappears in the Second Isaiah with more promises of good gifts and opportunities. A path in the desert with water, rivers and leafy trees is a beautiful memory to evoke so as to sustain people's hope.

Isaiah 43:16–21 evokes memories of the song of victory, sung and danced by Miriam and the women. This text is from a tradition of women's song and prophecy. The memories again move back to the exodus. There are also deep connections between women and water: women gather around wells, which are places of resistance, popular places for meetings and dialogues.

Today, as in biblical times, water is one of the most important issues in the world. For example, twenty percent of the Brazilian population (about 37 million people) do not have access to potable water; in rural areas, ninety percent of the population live without proper sanitation, including access to clean drinking water. The crisis has reached the peripheries of the cities. Basically, it is the poor who remain thirsty. Yet, water is a fundamental source of life. Through the reality of water, we can approach social, economic, political and religious realities. Because of this sense of integration, theology can be done from the margin of a river, at the border of a fountain or a well.

In the Bible we read about people who experienced being without water. From out of their daily experience with water scarcity, a whole theology of water has developed. For instance, paying for access and use of water was synonymous with oppression and a lack of freedom (Deut 2:27–28: "You shall sell me food for money, so that I may eat, and supply me water for money, so that I may drink").

The popular memory was kept alive through songs and stories. The wells were made by princes and nobles (Num 21:17–18: "... the well that the leaders sank, that the nobles of the people dug ..."). Yet wells are very important places, places of blessings, as we read in Genesis 21:30–31, as well as places of hope and promise of life, such as for Hagar and her son in the desert (Gen 21:19). Cisterns, wells and reservoirs are associated with abundance and blessings. Around them are vineyards and olive trees (Deut 6:10–11); celebrations of peace (2 Kings 18:31 and Isa 36:16); God's goodness is present with food and drink (Neh 9:25; Eccl 2:4–6).

In the New Testament, there is the wonderful dialogue between Jesus and the woman from Samaria (Jn 4). Here, around a well, the text describes a theological debate between Jesus and the Samaritan woman. They discuss daily needs such as bread, hunger, water, thirst, conflicts and desires. The meeting takes place in the woman's land, Samaria. Jesus moves from his land, his boundaries, his ground and crosses into the foreign land. In this way, the text makes obvious who lives there. When paths are already well-known, they can be walked without thinking about them. The theological discussion in this text mixes daily life with its pain and pleasures, concerns and needs, with people's vision and experience of God. The well is a site for this kind of discussion. The theological dialogue here is inspired by space. It is a theology from the margins, around the well.

In several biblical stories, the well is a point of encounter. It creates villages, oases and communities around itself. Around the well, news circulates and happiness and sadness are shared. Jacob's old well, a witness of many loves, fights and friendships (Genesis 29) is also the scenario of the meeting between Jesus and the Samaritan woman, who bears witness to the revelation of Jesus Christ as the Messiah.

Jesus was tired and thirsty. The disciples had gone into the town to buy food, but maybe had not understood what was really needed. The encounter evokes a reflection about what is needed, what is desired—what Jesus needs, what the woman needs and desires; what is necessary for drawing the water from the well. Jesus offers living water but he has nothing with which to draw the water. The woman has the practical tool and the knowledge of how to get the water. The well is a space for exchanging knowledge. At the well, it is possible to talk and share thirsts and lives.

The theology that is proposed here is a theology on the road, in process. It is a theology in movement. Through this movement, space and the sacred are reconfigured. It is a changing theology in times of climate change. It is a theology which does not fear changes, if they are guided by justice. The contextual reality is the starting point. Experiences from marginal people who suffer the consequences of climate change are central for discussing theology and reading Bible texts.

Third movement: climate change challenges us to reconfigure sacred spaces

Reconfiguring the sacredness of space requires attention to how we do theology (methodology). In times of climate change, theology also changes. We need to adapt, redraw and reconfigure how we understand theology—as movement, in transition and changeable.

We need to reimage God, creatively contextualizing the sacred and articulating how our experiences are interlinked with the sacred. Two biblical texts inspire this. In Job 38 and 39, God answers Job's arguments and questions out of the storm with a beautiful explanation. The text is a careful description of the places and functions of all beings created by God and their conviviality in nature. The image of God emerging from this text is very familiar and close to creation. God knows the creatures' needs and desires. God's wisdom is rooted in the harmony of life.

The other text is Deuteronomy 32:1–4. In the song of Moses, we can hear a blessing of eco-theological wisdom. Again, the experience of God is connected with and felt in the rain or dew. To express the experience of learning the Word of God, the image of the rain descending on the grass is needed. In the Old Testament, images from nature are commonly used to express the experience of God. The song expresses this knowledge in a poetic way.

The other space for re-envisioning sacredness is where theology is cooked or baked. Food sovereignty is a crucial concept in the reflection on climate change. When discussing climate change in connection with food sovereignty, as proposed by the small farmers or landless women's movement in Brazil, the space for reflecting and doing theology must change.

Feminists affirm that private life is political and permeated with power. Reformatting sacred spaces allows fresh air to blow into fixed, old structures, and makes changes in theology possible. This opens the windows for biodiversity in theological construction. In the words of Ivone Gebara,

> Religious biodiversity gives a heartfelt welcome to the diversity of tapestries. It is an exercise of going beyond our striving to make one single group the herald of a one-and-only truth, the self-appointed bearer of salvific formulas for everybody. Religious biodiversity implies an attitude of humility, which means that there cannot be absolute power that regulates and dictates the meaning of life, or the art to weaving meaning, or the art of evoking presence that are dear to us.[7]

This is the preface for a theology in pilgrimage, contextualized and in dialogue with the biodiversity of the cosmos and the earth. Theology is challenged by climate change prophetically and boldly to reconfigure space.

---

[7] Ivone Gebara, *Teologia Ecofeminista* (São Paulo: Olho d'Água, 1997), p. 108.

Listen to the Voice of Nature. Indigenous Perspectives[1]

Tore Johnsen

Introduction

The spiritual traditions of Indigenous Peoples have an important contribution to make in the search for a vision that can nurture a faithful Christian response to the ecological crisis. In the midst of climate change, can we discern a voice, a message about our own place in the mysterious gift of life? "Listen to the voice of nature!" an elderly reindeer herder from my own people, the indigenous Sami,[2] once told me. He had learned it from the old people in the *siida*[3] where he grew up.

> The voice of nature is like if you take too much, then nature tells you that those forests do not grow anymore, the reindeer will vanish, or the fish in the lakes disappear. ... You have to listen to the voice of nature. You are not deciding the future, but nature. You will not live in the future. Coming generations need the same nature to live off as you. You are supposed to leave it in the same condition as you have received. This were the admonitions concerning the future that I received both from my grandfather, my father an others in my *siida*.[4]

Yet another example from Sami tradition reflects the idea that creation has a voice that should be listened to. According to a story, in which

---

[1] This essay is based on a previous essay, "Teologi fra Livets Sirkel: Økoteologiske refleksjoner med utgangspunkt i samisk joikepoesi og indiansk filosofi," published in Bård Mæland and Tom Sverre Tomren (eds), *Økoteologi kontekstuelle perspektiver på miljø og teologi* (Trondheim: Tapir akademisk forlag), pp. 211–28.

[2] The Sami are the Indigenous People of northern Scandinavia and the Kola peninsula in Russia. For many the Sami are known as "Laplanders," but this word, which derives from a Scandinavian word, has negative connotations.

[3] The Sami word *siida* refers to a group of nomadic reindeer herding families who migrate every year between the summer lands in the coastal mountains and the winter lands in the interior mountain plateau.

[4] Author's own translation from the North Sami.

indigenous spirituality has obviously been blended with Christian thought, everything had the ability to speak in the distant past. On the Day of Judgment, every creature will once again be able to speak, first the dog and afterwards every other creature. Therefore, we should be careful in how we treat out fellow creatures.[5]

The ecological crisis carries in itself a disturbing message: certain ideologies, philosophies and theologies have failed. Here I shall turn to indigenous traditions to assist us in interpreting the Bible in ways that are more attentive to the voice of the earth. Using an old text from my own people, I shall show that Christian creation theology is not only about biblical texts, but also about who has to power to interpret, in whose interest, and through whose cultural lens? After establishing this critical awareness, inspired by the rich religious philosophy of Native North Americans, I will explore how Christian theology can move from an anthropocentric to an ecological paradigm,[6] according to which the human being is understood within the wholeness of creation. Although the focus here is on the ecological dimension, this is also intertwined with fundamental justice issues that climate change poses.

The thief and the shaman: alternative perspectives on creation

From an indigenous perspective, the theological tradition of the church is not necessarily a part of the solution to the environmental crisis. It can also be seen as part of the problem. Over the last centuries, in the course of Christianization, Indigenous Peoples have seen that the church has often felt that it had little or nothing to learn from spiritual traditions that emphasized an ecological vision of life. Rather, indigenous spiritual traditions were met with condemnation, demonization and persecution. In addition, Indigenous Peoples experienced that Christian creation theology could serve political interests and was used as an ideological basis for colonization and the exploitation of the land.[7]

---

[5] Johan Turi, *Mui'talus sámiid birra* (Uppsala: Almqvist & Wiksells, 1965), pp. 68–69.

[6] The phenomenon of theological paradigm shifts within the history of theology is well documented by Bosch for instance who has documented this related to the understanding of mission. See David J. Bosch, *Transforming Mission: Paradigm Shifts in Theology and Mission* (Maryknoll: Orbis Books, 2004).

[7] George Tinker, *Missionary Conquest: The Gospel and Native American Cultural Genocide* (Minneapolis: Augsburg Fortress, 1993).

The text of an old North Sami *yoik*,[8] put down in writing in the 1820s by the Finnish Lutheran minister Jakob Fellmann, testifies to the fact that the indigenous Sami also experienced colonization in this way. The *yoik*, entitled *Suola ja noaidi* (The Thief and the Shaman),[9] is in the form of a dialogue between "the thief," representing the colonizing culture, and "the shaman," representing the Sami.

From an eco-theological perspective, particularly the first part of the *yoik* is interesting. Here, the encounter between the thief and the shaman can be interpreted as a discussion about the moral basis for colonization. The *yoik* seems consciously to place the discussion within the context of the biblical creation texts in Genesis 1–3. The *yoik* starts with the thief saying: "My God is wandering." This hints at the paradise account in Genesis 3, where God is walking in the Garden of Eden (Gen 3:8).

The thief:
My God is wandering.
I have taken the plants of the earth
Grass and berries I have gathered
Trees and stones I have utilized.
I have not taken the property of the settled
I still take what grows on the land
A man came to me and called me a thief.

The shaman:
You do not know the way of the land.
Don't you know that I exist?
Look at the plants and take care
Observe the signs in the trees
Look differently at the grass.

The thief:
Who are you?
Are you not a human being?
As if you are God.
Have you created the grass?

---

[8] *Yoik* (from the Sami word "juoigat," to yoik) is the traditional way of Sami chanting or singing.

[9] The original North Sami text and an interpretation of this are found in Harald Gaski, *Med ord skal tyvene fordrives: om samenes episk poetiske diktning* (Karasjok: Davvi Girji, 1993), pp. 46–50.

> Have you made the trees?
> Aren't you made of the ashes of the earth?
> Like an insect you creep like I.
> The grass is not yours
> Neither are the trees and the stones.
> You are lord over your property.
> What is good is good.
> I know that also you exist
> Black shaman on the earth.
> Stay and live at your place.
> Use the grass which you grow.[10]

This *Suola ja noaidi yoik* clearly shows two cultures' different conceptions and uses of the land. First, we read about how the land is used by the intruding culture. The words and expressions suggest that it consists of farming ("Taken the plants of the earth"), gathering ("Grass and berries I have gathered") and foresting and mining ("Trees and stones I have utilized"). In addition, property seems to be an important category in the mindset of the colonizing culture, and landownership linked with permanent settlement. The colonist thus says, "I have not taken the property of the settled." Therefore, he cannot understand why the shaman (the Sami) can refer to him as a thief. This indicates that the Sami were nomads and therefore, according to Western thinking, not entitled to property rights.

The Sami shaman in this *yoik* is an advocate for ecological values. He asks the thief to learn the way of the land, and to look differently at the trees and the grass. He should take care and look for the signs in the trees. In this way, he defends values and attitudes that we can find today in the modern ecological movement.

In this *yoik*, the thief's response to this challenge constitutes a critique of a certain use of Christian creation theology. The thief responds to the shaman with arguments drawn from Christian creation theology. At first, the thief's answer seems pious. He emphasizes that it is not human beings but God who created everything. However, if we examine the thief's line of argumentation more closely, we discover that his emphasis on God as Creator is not promoting ecological values. Instead, he uses the argument to promote colonization and to undermine the shaman,

---

[10] Author's own translation of the original Sami text.

his way of life and his ecological perspective. The thief is using creation theology to argue for a way of life where people own the land ("You are lord over your property"), are settled and not a nomads ("Stay and live at your place"), and are farming ("Use the grass which you grow"). This points to Genesis 2:15, where God places the human being to work the garden and to take care of it.

This theological reading of this *yoik* gives rise to an interesting observation. It seems that at the beginning of the nineteenth century, the Sami not only knew the biblical creation account, but some also recognized a connection between this creation theology and the colonization they had experienced. They realized that Christian creation theology had become an instrument for colonizing them and exploiting their land. It provided an argument for favoring a settled culture that emphasized landownership, and an argument against a Sami way of life that was close to nature. Although this text comes from long before the serious consequences of industrialization had become obvious, it clearly captures certain negative tendencies in Christian creation theology. It was not until one and a half centuries later that serious critiques of these tendencies appeared in Western thought.[11] The traditions and experiences of Indigenous Peoples provided a much earlier, important corrective to a common Western interpretation of Christian theology.

> Look at the plants and take care
> Observe the signs in the trees
> Look differently at the grass.

How can these words be interpreted theologically? They could be read as an attack on Christian creation theology as such, but might also be regarded as an alternative interpretation of Genesis 2, in which the communal nature of creation is emphasized. This would resonate well with Sami understandings of how, in the beginning, everything could speak and how everything will be able to so again on the Day of Judgment. This story suggests a spiritual vision of creation as a community, not an "empty" object. Creation has a spiritual dimension and consists of subjects with their own dignity before their Creator.

---

[11] Cf. Lynn Townsend White, Jr, "The Historical Roots of Our Ecological Crisis," in *Science*, vol. 155, no. 3767 (10 March 1967), pp. 1203–7.

The significance of the theological starting point

Christian theology strongly emphasizes the inviolable value of every human being. However, historically some interpretations have contributed to a weakening of ecological consciousness. Valuing human beings has sometimes been at the cost of devaluing the rest of creation. This tendency culminated at the time of the European Renaissance, when an anthropocentric worldview elevated human beings above the rest of creation. This paved the way for an instrumental understanding of nature, reducing creation to an instrument to serve humans. The self-realization of the human being became the perspective from which the rest of the world was observed and valued.

Faced with the effects of human-caused climate change, it now seems obvious that in ecological terms this anthropocentric worldview has failed. Through a conscious shift in the theological starting point, the ecological aspects of the Christian faith need to be further developed.

This is similar to the conscious shift that occurred in liberation theology, "doing theology with a preferential option for the poor."[12] God is not indifferent to injustice. God always takes sides. Since God stands at the side of the poor and oppressed, suffering and struggling with them, liberation theologians have advocated that we need to make conscious choices as to our theological starting point. The validity of this shift in theological starting points is now widely recognized in theological circles, and is a major contribution to Christian theology in general.

I suggest that a similar shift in theological starting point is necessary today with respect to creation. We need to begin with creation as the place from which we begin our theology. Here the spiritual traditions of Indigenous Peoples can contribute to the needed reorientation of Christian theology.

Theology from the circle of life

I propose an eco-theological alternative to the idea of beginning theology from the option for the poor: doing theology from the circle of life.

---

[12] A discussion about the importance of the theological starting point is found in Sturla Stålsett, "Hvor finner teologien sted?," in *Norsk tidsskrift for misjon 95* (2) (1995), pp. 83–100.

What is presented here is only a suggestive sketch, intended to invite further dialogue and development.

"The circle of life" belongs to the religious philosophy of the Indigenous Peoples of North America.[13] It is articulated in the so-called medicine wheel philosophy of the First Nations on the Great Plains. The circle symbolizes the holistic nature of reality. We all belong to a great community where all things are interconnected and interdependent. The circle symbolizes the fundamental harmony and wholeness that exists between all created things. In this circle, the community is characterized by solidarity, reciprocity, respect and love. This fundamental solidarity not only embraces the human sphere of life, but reaches out to everything that exists: animals and birds, trees and stones.

According to this philosophy, the creator—Wakan Tanka, Kitse Manitou, the Great Spirit—is present in all things, and therefore present in the whole circle. At the same time, the Great Spirit is first and foremost understood as the very centre of the circle, the source of life that embraces all things. It is in the Great Spirit that the circle is whole.

Formulated in terms of Christian theology, we can say that the presence of the Great Spirit in the circle represents God's immanence in the created world, while the Great Spirit as the centre of the circle represents God's transcendence. The Lakota Sioux medicine man and catechist, Black Elk (1868–1950), expressed this theological paradox in this way:

> We should understand well that all things are the works of the Great Spirit. We should know that He is within all things: the trees, the grasses, the rivers, the mountains, and all the four-legged animals, and the winged peoples; and even more important, we should understand that He is also above these things and peoples.[14]

---

[13] For an elementary introduction to this philosophy, see Stanley J Mckay, "An Aboriginal Christian Perspective on the Integrity of Creation," in James Treat, *Native and Christian: Indigenous Voices on Religious Identity in the United States and Canada* (New York and London: Routledge, 1969), pp. 51–55, and George Tinker, *Missionary Conquest: The Gospel and Native American Cultural Genocide* (Minneapolis: Augsburg Fortress, 1993). For a more detailed introduction, see for example Black Elk's account about the seven sacred rites of the Oglala Sioux, in Joseph E. Brown, *The Sacred Pipe: Black Elk's Account of the Seven Rites of the Oglala Sioux* (Norman and London: University of Oklahoma Press, 1989).

[14] Brown, *ibid.*, p. xx.

Even though this expression affirms the presence of God in all things, it is still not proper to call this religious tradition "pantheistic." God is not reduced to the totality of all things. The Great Spirit is "also above these things and peoples." A more proper term for describing this spiritual tradition is therefore "panentheism," which means that everything is in God.

The medicine wheel philosophy is not completely harmonious, even though the circle is the fundamental symbol. Human beings are the only creatures who do not naturally find their place in the circle of life, and therefore have to learn how to find their place in creation and how to share their lives in solidarity with the rest of creation. To do so, they have to search for spiritual guidance and assistance outside themselves.

The weakness in human nature means that humans often fail. This causes brokenness in the circle of life. To mend the circle, to make it whole, therefore, is at the heart of this spirituality. This is reflected in the traditional ceremonies, which help humans to find their place in the wholeness of creation, to reestablish reciprocity and balance and to renew their relations with all other aspects of life. Restoring the wholeness of the individual is thus always understood in the context of restoring the wholeness of creation. My wholeness, the wholeness of the community and the wholeness of creation are only different dimensions of the same reality. This, in short, is the religious philosophy of the circle of life.

Theological concepts interpreted from the circle of life

In order to explore the potential of the circle of life to serve as a staring point for Christian theology, I will reconceptualize some basic theological concepts in light of the circle.

The circle represents the ideal condition of the entire creation. Even though God is placed at the centre of the circle, it is important to remember that the entire circle is filled with the presence of God. Furthermore, creation is not separate from the human sphere of life, but is the very context and essence of all human life. We are creation. We are earth beings, *adam of adama*, creations taken from the soil (cf. Gen 2:7). We exist by participating in the circle of life (cf. Ps 104).[15] It is

---

[15] In the spiritual traditions of many Indigenous Peoples the interconnectedness with creation manifested in the concept of mother earth. Modern biblical research shows that a similar complex of ideas underlies several Old Testament texts (i.e., Job 1:21; Eccl 5:14; Ps 139:13–16). Terje

this wholeness of creation that is expressed in Genesis 1:31: "God saw everything that he had made, and indeed, it was very good."

The wholeness of creation is broken by human sin. In Christian theology, the concept of sin is often conceptualized in a very individualistic manner. If we conceptualize "sin" from the circle of life, we can say that "sin" is the human violation of the wholeness that exists in the circle of life. The entire circle rests in God. Hence, a broken circle expresses an injured relation to God. Human sin expresses itself through a broken circle—at the individual, collective and ecological levels.[16] A broken relation to God leads to broken relationships within the circle—and broken relationships in the circle lead to a broken relation to God. The two dimensions are intertwined and interdependent.

Theologically we can say that the circle (and its whole or broken condition) expresses our relationships both horizontally (toward our neighbors and fellow creatures) and vertically (towards God). In the circle, the two levels are held together. The reality of sin implies that we now live in a broken circle of life.

In the broken circle of life, God—the Creator, Sustainer and Redeemer of all things—is still at the centre of the circle. God's presence continues to penetrate creation. God is still the one who holds everything together. The fragmentation of the circle is only a consequence of the fact that humans dissociate themselves from the God of life. The wholeness of the circle flows from and rests in God.

But how does Christ fit into this model? First, Christ belongs to the center of the circle. As the Son of God, the preexistent *logos*, the cosmic Christ, he is a part of the deity. Christ is the revelation of the sacred source and centre of cosmos—what we call God. Understood in this way, Christ is a part of the source from which the circle of life emanates and in which it exists. This is consistent with the biblical witness about Christ (cf. Jn 1:1–4; Col 1:15–17).

Through the Christ event, Christ is rooted not only in the center of the circle (cf. Jn 1:9–14). Through the incarnation, Jesus Christ is united with the circle of life, all of creation. In Christ, God has not only become

---

Stordalen, "Moder jord'-en etisk impuls i Det gamle testamente," in Jan Olav Henriksen (ed.), *Makt Eiendom Rettferdighet: Bibelske moraltradisjoner i møte med vår tid* (Oslo: Gyldendal Akademisk, 2000), pp. 115–38.

[16] An indigenous interpretation of the Christian concept of sin hinting in this direction is found in Paul Schultz and George Tinker, "Rivers of Life: Native Spirituality for Native Churches," in Treat, *op. cit.* (note 13), pp. 56–67, here pp. 65–66.

a human being. God has become *sarx* (human flesh) (cf. Jn 1:14). God has become an earth being, a human being of the soil—the new "adam of adama" (cf. Gen 2:7).[17]

What about the cross, the suffering and death of Christ? Theologically speaking we can say that the Christ event takes place not only for the sake of humanity, but because of God's care for the whole world (cf. Jn 3:16). God has not only united Godself with a suffering humanity. Salvation embraces the entire suffering and broken creation (cf. Rom 8:19–22). In this way, the cross can be interpreted as the presence of Christ in the broken circle of life, opening up an eco-theological interpretation of the cross.

The reconciling act of Christ and the message of salvation are addressed to humans in a special way (cf. Jn 1:4) because it is through human sin that the circle is broken, and therefore only humans need to be liberated from sin (cf. Rom 5:12–19). In this sense, the focus of salvation is anthropocentric. It is addressed to humanity, the cause of disharmony in creation. However, humans are also key in bringing back wholeness to creation (cf. Rom 8:21). In this way, the message of salvation in Christ is not only anthropocentric but within an ecological horizon, its goal is the restoration of the entire creation (cf. Eph 1:10; Col 1:20)

Thus, the cross has a role to play, not only with regard to humans but also for the entire creation. The cross is liberated from a limited anthropocentric interpretation, and understood as expressing God's compassion and restoration of the broken creation. The presence of the cross—symbolizing the suffering and resurrected Christ—in the broken circle of life can thus be understood in two ways. First, it expresses God's solidarity with the broken world. Second, it expresses God's saving presence in the broken circle of life.[18] By the cross of Christ—through his suffering, death and resurrection—God is once again centering the universe toward and around Godself, and is thus making the circle whole again (cf. Eph 1:10; Col 1:20)

---

[17] The fact that the human being, adam in Hebrew, is taken from the land, is fortified by a play of words in Gen 2:7. The Hebrew word for "land," from which the human being is formed, is adama. The human being can therefore be called "adam of adama," an "earth being." Cf. Terje Stordalen, *Støv og livspust: Mennesket i Det gamle testamente* (Oslo: Universitetsforlaget, 1994), p. 72.

[18] An ecotheological interpretation of the cross related to a Sami tradition is found in Tore Johnsen, *Sámi luondduteologiija / Samisk naturteolog - på grunnlag av nålevende tradisjonsstoff og nedtegnede myter* (Stensilserie D nr. 04, Institutt for religionsvitenskap, Det samfunnsvitenskaplige fakultet, Universtitet i Tromsø, 2005), pp. 72–73.

What does this shift in the theological starting point imply for reconceiving other central theological concepts such as the kingdom of God?[19] In Christ, a new reality has broken into the world. This is called the kingdom of God, and is the eschatological hope for a fully restored circle of life—a new creation, a new heaven and a new earth (cf. Rev 21:1). This eschatological reality has already broken into the world, even though not fully. While it will be fully realized only in the future, it is however already present in Christ. The kingdom of God that is the restored circle of life, the reconciled creation, is a reality that can already be experienced now in a mystical way in Christ. As Luther expressed it:

> Christ [...] fills all things [...]. Christ is around us and in us in all places [...] he is present in all creatures, and I might find him in stone, in fire, in water, or even in a rope, for he certainly is there [...].[20]

This allows for an ecological Christ mysticism. The body of Christ can be interpreted as the mystical expression of the restored circle of life, suggesting an ecological interpretation of both Eucharist and ecclesiology.

If the eschatological reality brought near in Christ is the restored circle of life, then this is the reality we actually celebrate in the Eucharist. We might therefore call the Eucharist the liturgical expression of the restored circle of life. In Christ, the wholeness of life is restored—not only for the human family, but also for the entire creation.

In the same manner, an ecological interpretation of salvation calls for the development of an ecological ecclesiology—that is an understanding of the church which is defined and developed with the circle of life, or creation as its context. Interpreted from the circle of life, we might say that the church is to be an expression of the new humanity reconciled both with its Creator and its co-creation.

What would such an ecclesiology mean for the church in the world? This implies that the church's diakonia is not only in terms of care for

---

[19] An interpretation of the "kingdom of God" from Native American perspective is found in George Tinker, "Spirituality, Native American Personhood, Sovereignty, and Solidarity," in Treat, *op. cit.* (note 16), pp. 115–31, here pp. 125–27.

[20] Martin Luther, "The Sacrament of the Body and Blood of Christ—Against the Fanatics," in Timothy Lull (ed.), *Martin Luther's Basic Theological Writings* (Minneapolis: Fortress Press, 1989), p. 387.

humans, but also in service of a broken creation.[21] In terms of worship, liturgy should embody the circle of life as the very place from where we worship. We are united with God's creation as we worship our Creator and Savior, and we stand in solidarity with the broken creation as we pray for reconciliation and restoration in our lives. To enact the liturgy from the circle of life implies a conscious attempt to embody creation as the very place from where we worship.

Creation becomes an essential dimension not only of the two first articles of faith, but of the third article as well. Here the one holy, catholic and apostolic Church and the communion of saints are understood within the context of the wholeness of creation. This contributes to a reconciliation between two dimensions of pneumatology which normally are hard to hold together in theological terms, namely the work of the Spirit in creation (cf. Gen 1:2; Ps 104:30) and the work of the Spirit related to Christ and the church. In Western theology, normally only the latter is emphasized, while the work of the Spirit in creation receives little or no attention.

In conclusion, the moving from an anthropocentric to a more ecocentric theological paradigm is made possible by a theological model that is essentially theocentric, christocentric and trinitarian. Some Christians may fear that by paying more attention to creation we risk drifting away from the essentials our faith. The above suggests, however, that an ecological turn allows for a rich appropriation of the biblical witness.

Closing reflections

Some years ago, I heard an Aboriginal elder in Canada referring to a story about Jesus healing a blind man. In this story, Jesus takes soil from the earth and puts it on the eyes of the blind man (cf. Jn 9:1–7). Referring to how Jesus used the soil for healing, the elder said, "This is our medicine!"

---

[21] Line M. Skum, herself a Sami person, criticized the diakonia definition of Church of Norway as being anthropocentric, and argued for the integration of ecological perspectives in the diakonia concept. Line Merethe Skum, *Når moder jord trues: en studie av hvordan den episkopale kirke og United Church of Christ tenker og handler frigjørende sammen med Igorot i Mankayan i kampen om land som trues av multinasjonal gruvedrift. Hovedfagsoppgave i helsefag, studieretning diakoni* (Universitetet i Oslo, 2005). In 2007, the national synod of the Church of Norway (Kirkemøtet) passed a new plan for diakonia where the ecological dimension is included in the new definition: "Diakonia is the gospel in action expressed through love for the neighbor, inclusive community, protection of creation and struggle for justice." (Author's own translation.)

Is the current climate change crisis like this piece of earth smeared upon our eyes? In the midst of the despair and suffering caused by climate change, is there also a potential for healing and restoring our spiritual vision? Is the healing of our spiritual vision a matter of reconnecting with the soil we are taken from by the healing hands of Christ?

Christian theology needs to move from an anthropocentric to an ecological paradigm, drawing upon important perspectives from Indigenous Peoples. This involves a reorientation of our theological anthropology. We need to start to understand and experience ourselves as taken from the earth and woven into the great web of life. It was as a piece of soil that the human being received the breath of God. This must imply that it is as a "piece of soil" and within the context of creation that we will realize our calling as the image of God. To understand and live our lives from the circle of life calls for a new self-understanding, an ecologically conditioned anthropology. The earth becomes part of our identity.

> Look at the plants and take care.
> Observe the signs in the trees.
> Look differently at the grass.
> Listen to the voice of nature!

Wisdom Cosmology and Climate Change

Norman Habel

While standing on the shore near Puri in Orissa on the Bay of Bengal, where the water is washing away the land, I wondered about the words of wisdom in Proverbs, "[God] assigned the sea its limit, so that the waters might not transgress his command (Prov 8:29).

Thinkers around the globe are currently trying to understand the nature of climate change and how we can best come to terms with the crisis that seems imminent. We could view this crisis as God's judgment on the greedy ways of humanity—a prophetic approach. We could see it as but another cycle in the long-term weather patterns that surround our planet—a paleo-climatology approach. Or, we could perhaps read it as a God-given sign of the end times—an apocalyptic approach.

There is, I believe, a way of approaching this phenomenon that has been given little consideration by biblical scholars. Within the biblical traditions, wisdom represents a school of thought that seeks to understand both society and nature in a realistic way. Observing nature is an integral part of the task of "gaining wisdom." And gaining wisdom, I suggest, may be of considerable value in the current environmental climate.

To explore this option, I think of the wisdom school in the Old Testament in terms of a coherent cosmology. This wisdom cosmology is to be distinguished from cosmologies reflected in other biblical traditions. In many ways, the wisdom cosmology is more consistent with current ecological worldviews and reflects an approach to cosmology that is grounded in observation and analysis by the wise, rather than theological reflection based on ancient motifs or myths. The wise may be classified as the scientists of the ancient world, sages who depend on observing creation and society rather than direct revelation from God.

A wisdom cosmology

The "way" of things

Each phenomenon and domain of nature has an innate code (*derek*) or law that governs its characteristic behavior as an integral part of the cosmos.

Have you ever wondered why a frog always jumps like a frog and never runs like an ant? Have you ever been fascinated by the way in which a baby bird learns to fly as a bird rather than swim like an eel? There is something inbuilt in each creature that enables it to be true to its nature.

Some scientists have examined this phenomenon and sought to explain it in terms of genetics or ecosystems. Each creature, it seems, has an inbuilt genetic code that governs how it lives, moves and reproduces. Quite remarkably, the differences between the genetic coding in a human being and that of any other animal are relatively few. Yet, these inner codes determine how a snake moves in a way quite distinct from that of a snail or an eagle or a human!

This mystery is also one that occupied the minds of certain wise individuals in the ancient world. The wise used a number of terms in reference to this inner code. The most explicit term is "the way" (*derek* in Hebrew). Because "the way" has a wide range of meanings in English, the technical meaning of this term is easily overlooked in translation. We can speak of a way of doing things without having a specific technical *modus operandi* in mind. And in wisdom literature the term can also be used generally in reference to the ways of the wicked or the ways of the righteous.

The term "way" (*derek*), however, also has a technical sense.[1] It refers to the inner code of behavior that characterizes a phenomenon of the natural world. The way of something reflects its essential character, its instinctual nature, its internal impulse. So the way of an eagle is to soar across the sky and with its eagle eye to discern prey far below. The way of a snake is to slither across rocks without any legs and to camouflage its presence in the grass.

In this usage, *derek* refers to the driving characteristic of a given entity. The German equivalent used to classify creatures is the term

---

[1] Norman Habel, "The Implications of God Discovering Wisdom in Earth," in Ellen van Wolde (ed.), *Job 28: Cognition in Context* (Leiden: Brill, 2003), pp. 281–98, here p. 286.

bildende Kraft (forming force). Every phenomenon of nature has its *bildende Kraft*, its own inner formative force, or as I suggest, its driving characteristic.[2] This concept of *derek* is crucial for an understanding of wisdom cosmology.

There is a measure of mystery in the "way" of things, as one of the ancient wise ones states: "Three things are too wonderful for me; four I do not understand: the way of an eagle in the sky, the way of a snake on a rock, the way of a ship on the high seas, and the way of a man with a girl" (Prov 30:18–19).

It is the task of the person interested in becoming wise to observe nature and understand its "ways" closely. In so doing, the individual may not only learn about the phenomenon itself, but also learn lessons about life. The novice is encouraged to watch the ant, observe the distinctive code of behavior found in ants and so to gain some wisdom: "Go to the ant, you lazybones; consider its ways, and be wise. Without having any chief or officer or ruler, it prepares its food in summer, and gathers its sustenance in harvest" (Prov 6:6–8).

Here the wise teacher is leading a youth to analyze the "way" or "ways" of an ant. What characterizes an ant? What makes an ant an ant? What is the code or "way" of an ant? If you figure out that mystery, you have gained some wisdom!

The way or code that is typical of the nature of ants, according to the wise, is for them to function as a corporate body without any hierarchy, without any bosses or leaders, a mystery that modern scientists still find fascinating. Another code that is true of ants is their instinctual capacity to gather food in summer and store it for the winter.

As noted above, the wise are the scientists of old, those committed to observing phenomena and trying to "discern" their very nature. "To discern" *(bin)* might readily be translated as "to research through close examination." The person with the necessary cognitive skills of discernment can discover the "code" or "way" of what is being examined. Such is the task of the wise, a task we may want to emulate in the current climate context, especially when we see species disappearing and their "way" on our planet becoming extinct.

---

[2] Umberto Eco, *Kant and the Platypus. Essays on Language and Cognition* (London: Vintage, 2000), p. 93.

The "ways" of the climate

Climate is a web of connected domains in nature, each with its distinctive code and locus in the cosmos.

The preceding examples belong to the animal world and represent biological ways, but what about the climate? The wise men of the Old Testament explored the way or code of every part of creation, both animate and inanimate. The weather or climate was no exception. Job 38 is essentially an exploration of the codes in many parts of creation. Various terms are used to describe the code or basic nature of something—way, limit, law and wisdom.

The exploration of the designs of creation where these codes function is often given in very poetic language—in a series of portraits. Yet, in each portrait, there are terms that point to the code that God challenges the wise to find. In this case Job is the science student.[3]

The initial challenge addressed to Job is not to obscure the "design" (*'etsa*) of the universe with all its mysterious wonders and ways. Here design is to be understood as the structure of the cosmos founded on wisdom, a deep underlying mystery that governs nature. As such, it is to be distinguished from the modern concept of "intelligent design." Job—and those of us who would be wise—are invited to explore the cosmos so as to comprehend the codes that govern its operation.

In this exploration of the cosmos, God challenges Job to locate the "way" of light (Job 38:19) and the "way" of the thunderbolt (Job 38:25). Especially significant is God's portrayal of the various constellations in the sky (cf. Job 38:31-33). Not only is Job challenged to understand the code of these constellations but also to understand the "laws" of the skies (Job 38:33). The skies above, with all their galaxies and constellations, also have an inner code which is here called the "law of the skies." God even challenges Job to try and take the laws of the skies and establish them on earth. Impossible! Each domain has its own code and its own locus (*maqom*) in the cosmos.

God also challenges Job to consider the clouds in the sky. They are not random movements above earth. Instead each cloud possesses an inner code, or as God says, "Who put wisdom in the cloud canopy? Who gave discernment to my pavilion?" (Job 38:36, author's own translation).

---

[3] Habel, *op. cit.* (note 1), pp. 287-89.

According to the wise person who wrote Job, in the design of creation, each domain of the physical universe has inbuilt codes or laws designed to function in a way that is true to its created nature. Clouds do what clouds do because of the innate wisdom they possess.

The "ways" of thunderclouds and lightning are viewed as central to the weather patterns that bring rain on land "where no one lives" and stimulates the soil to produce grasses and trees (Job 38:25–27). Job is challenged to understand the ways of the weather, the codes in the climate. To understand nature, the wise must explore that mystery and treat it as such.

While we may be mesmerized by meteorologists who seem to answer God's challenge and explain the mysteries of the weather, it is clear that we have been abusing the atmosphere and not discerning its way. Human wisdom requires that we both understand the way of the atmosphere and work in harmony with it. The physical universe is a complex of components, each of which has a distinctive "law," "way," "place" or "dimension" that characterizes it in relation to the rest of the cosmos.

The place of wisdom

Wisdom (*kochma*) is the underlying design of the cosmos, a dynamic blueprint that orders, governs and integrates the various codes in the domains of the cosmos.

In the continuing process of creating domains of the cosmos and establishing their respective codes, God discerned wisdom as the integrating blueprint, the cosmic code that governs all codes.

While in one text in Job 38 wisdom is equivalent to the code itself in the clouds (38:36), there are texts which indicate that wisdom is really the code behind all codes, the designing impulse, the dynamic blueprint for the various "ways" or codes in the domains of creation.

Job 28 raises the ancient question, Where can wisdom be found? Humans may probe the depths of earth for precious metals and other hidden things, but the locus or "place" of wisdom in the cosmos remains a mystery. "Place" (*maqom*) is a technical expression referring to the locus of a phenomenon or domain in a given ecosystem of the cosmic order. The ultimate question is the "place" of wisdom in the cosmos. Clearly, wisdom is here not a body of past knowledge or a deep level of understanding. Wisdom is not the accumulated tradition of the wise elders. Wisdom is a prior reality in a mysterious "place."

The hidden dimension of wisdom is apparent to all. The deep claims "it is not in me," the birds of the air cannot detect it; the land of death has only heard a rumor of it. God is the one who knows its "place." But God's knowledge of the place of wisdom is not some innate divine capacity. God, like a true scientist of old, observes all the phenomena on earth, examines all the domains under the sky and in the process discerns the locus of wisdom in the cosmos. "God understands the way to it, and he knows its place. For he looks to the ends of the earth, and sees everything under the heavens" (Job 28:23–24).

And where does God specifically look in God's search of the realms under the sky? In the domains of nature we would connect with climate—the wind, the waters, the storm and the rain. And, significantly, God makes this close examination when creating these various domains and establishing their nature. So wisdom is connected with the process of creation from the beginning.

Just as significant here is that God does not simply observe these domains of nature, but focuses on their several codes described variously as the force/weight of the wind, the volume/measure of the waters, the rule for the rain and the way of the thunderstorm. The codes of these domains of creation provide the clue to something more, something deeper, something all encompassing, namely, wisdom. These codes of creation connect with the hidden design, the blueprint that determines their nature. Behind all the codes of the climate is a dynamic blueprint called wisdom.

> When God fixed the weight of the wind and meted out the waters by measure, when God made a rule for the rain and the way of the thunderstorm, then God saw her (wisdom) and appraised her, established her and probed her (Job 28:25–27, author's own translation).

When God discovers this hidden mystery at the core of creation, the divine Sage checks and double checks the findings. They are appraised and probed. In the process, wisdom is "established" as the ultimate design that governs the codes of creation. Wisdom is not an attribute of God's but a deep dimension of the cosmos connected with the codes in creation. Wisdom is the code that integrates the various codes in nature, the ultimate ecosystem that also governs the weather patterns of the world.

"God's way before God's works"

Wisdom is also viewed as having a special personal relationship with God, a living presence who, from the very beginning, guides God in the creation of the various domains of the cosmos and the establishment of their respective codes.

The preceding interpretation of Job 28 provides a basis for reexamining the famous wisdom poem of Proverbs 8:22–31, where wisdom presents herself as a companion of God at the primordial time of creation. Crucial for an appreciation of God's relationship with wisdom is how the opening verse of this poem is rendered. I favor here a literal translation as the most enlightening.[4]

> YHWH acquired me first, his way before his works. ... from of old, from antiquity I was established, from the first, from the beginnings of earth (Prov 8:22–23, author's own translation).

The verb *qana*, here translated "acquired," is the standard term employed in Proverbs for acquiring wisdom (as in 4:3, 7). The repeated injunction of the teacher is to "acquire wisdom." Given this, it seems logical to understand this term in the standard way rather than to seek a relatively obscure meaning of "create" from other contexts such as Genesis 14:19. Moreover, the precedent of Job 28 makes it clear that God, as the Sage, may be understood as discovering wisdom as an existing or emerging reality in the world. It is also evident from Proverbs 8:24–25 that wisdom "emerges" prior to creation. She is not created but embraced by God in some way.

Also important is the translation of *derek* in its technical sense of "way" or code. To render *derek* by the term "work" (as in the NRSV), is to conceal the differentiation between the "way/code" of something and the making of something. Wisdom claims to be the "way" that precedes the work, the creative activity of God. This priority of wisdom as the way is apparent throughout the poem. Wisdom is not God's "work," but a prior presence.

What then is the function of wisdom as the "way" that precedes and accompanies God at creation? Clearly, wisdom is more than a friendly companion. As in Job 28, a clue may lie in how the various domains of

---
[4] *Ibid.*, p. 294.

creation are described. In some verses, the focus seems to be on the domain itself. In others, however, wisdom announces her presence prior to that dimension of a domain, which we might again call its code, its governing characteristic. Wisdom, like the blueprint in Job 28, seems to be the dynamic design that inspires God and determines the various codes of creation. As the "way," wisdom is the blueprint that provides the codes for creating the various domains of creation. The design precedes the act of creation, the way precedes the works of the Creator.

The personification of this "way" indicates that wisdom also has a dynamic personal relationship with God. Wisdom inspires and informs God in the creation process. For God to "acquire" this way at the very beginning is to recognize God as the supreme Sage. It is this Sage who employs wisdom as the means of creating the cosmos: "The Lord by wisdom founded the earth; by understanding he established the heavens" (Prov 3:19).

Another conundrum occurs in Proverbs 8:30 where the rare term `amon appears. To make the translation somewhat logical, interpreters have rendered it as "master craftsman," suggesting that wisdom was God's primordial architect. Willilam McKane renders the term "confidant" and makes a possible association with the queen mother concept as in an Egyptian court. As McKane indicates, the Massoretic text reading seems to be that of a "darling" child.[5] If so, then the text reflects a fascinating irony.

Whatever the primary metaphor, it is clear that the "way" that God has acquired and has informed God in creating the cosmos is personified as a presence that God discovers, embraces and celebrates. To highlight God's delight in this relationship and process, wisdom seems to be described as a "darling one" beside the Maker.

It is not insignificant that the closing announcement of wisdom is one of celebration. Wisdom rejoices in creation and with humanity. The living design that integrates the codes of the cosmos is not viewed as a passive, lifeless blueprint, but a vibrant, living presence. The mystery called wisdom at the core of the cosmos is a voice inviting all who would be wise to explore her worlds, discover her presence and celebrate her codes. Or as the teacher enjoins, "Say to wisdom, 'You are my sister,' and call insight your intimate friend" (Prov 7:4).

---

[5] William McKane, *Proverbs a New Approach*, The Old Testament Library (Philadelphia: Westminster Press, 1970), p. 357.

Acquiring wisdom

Human wisdom involves closely observing the phenomena of nature, discerning the respective codes within each, exploring their interconnection with wisdom as the integrating design of the cosmos and learning to live in harmony with these phenomena and the mystery called wisdom that governs them.

Just as the divine Sage acquired wisdom in the primordial (Prov 8) and in the creation of the weather system (Job 28), humans are encouraged to "acquire wisdom." In Job 39, God takes Job on a tour of the animal kingdom to help him understand the challenge involved. For Job to acquire wisdom he is summoned to answer questions about the creatures of the wild to demonstrate whether he understands them and their "ways." Job had a right to challenge the cruel sufferings that God had imposed on him, but surprisingly God responds by challenging Job's understanding of the mysteries in the natural world.

Job has several important lessons to learn. The first is that the ways of the creatures in the wild are complex and mysterious. There is more to the kingdom of the wild than a random array of strange animals with odd antics. Each creature has its way, its distinctive code of behavior, its innate character. God challenges Job's capacity to grasp the mysteries of pregnancy and birth among creatures like the ibex, which have no midwives to help them. God challenges Job's ability to find food for young lions waiting in their lairs or young ravens crying out to God. And God challenges Job's capacity to comprehend the "way" of the eagle soaring through the sky to identify its prey far below.

This crucial lesson in wisdom also relates to Job's understanding of his relationship to nature. Essentially God is saying to Job that human wisdom means more than knowledge; it also means that human beings do not have control or dominion over such animals. They are wild and free; that is their very nature. Humans are not commissioned to domesticate or dominate them. This challenge is reflected in a delicious satire in which God asks Job to demonstrate that he can get a wild ox to "serve" him, to babysit beside his crib, to do his farm work for him and finally to bring in the harvest. The text reads:

> Is the wild ox willing to serve you? Will it spend the night at your crib? Can you tie it in the furrow with ropes, or will it harrow the valleys after you? Will you depend on it because its strength is great, and will you

hand over your labor to it? Do you have faith in it that it will return, and bring your grain to your threshing floor? (Job 39:9–12).

This wisdom speech seems to reflect a tradition which challenges the mandate to dominate extended to humans in Genesis 1:26–28. Not only do humans have a mission to serve creation in Genesis 2:15, they also have an invitation, in passages such as Job 39, to understand the "ways" or ecosystems of the natural world, respect them as they are and live in harmony with them.

Job is forced to realize that as a human being who aspires to be wise, his mission is to discern not to dominate; his mandate is to explore the mysteries of the natural world rather than to exploit them.[6] God calls Job to listen, respect and be humble before nature not to be arrogant. Acquiring wisdom means observing, discerning and realizing that every "way" in nature is ultimately connected to that deep mysterious impulse in the cosmos called wisdom.

Pursuing folly

Human folly means seeking to dominate rather than to discern the respective codes in various domains of the cosmos and daring to violate these codes in such a way so as to upset the balance in the laws of nature.

The folly of humanity over the last hundred years or so, in terms of a wisdom cosmology, is immediately apparent. Humans have been bent on dominating the many domains of nature, especially arable land, forests and rivers. The domains of nature have been viewed as God-given resources to exploit at will. The environmental implications of this exploitation are evidence of our folly.

Especially significant is the folly involved in breaking the boundaries of several domains. Each domain has its "place" and the elements of each domain have their "place" or locus in nature. Humans, however, have extracted fossil fuels from their "place" in the domains below the earth, where they belong, transformed them into various gaseous forms and disseminated them in another domain, namely, the atmosphere. The

---

[6] Norman Habel, "'Is the Wild Ox Willing to Serve You?' Challenging the Mandate to Dominate," in Norman Habel and Shirley Wurst (eds), *The Earth Story in Wisdom Traditions. Earth Bible Volume 3* (Sheffield: Sheffield Press, 2001), pp. 179–89, here p. 180.

result is an atmosphere overloaded with greenhouse gases and a disruption of the existing balance in the "ways" or laws of nature.

The dilemma we now face is a change in climate patterns. The previous codes which governed the cycle and pattern of the winds, seas, storms and droughts seem to have been disrupted. The laws that govern the weather patterns seem to have changed.

I illustrate from the 2009 bushfires in the State of Victoria in Australia. As a boy on the farm, I knew the "way" of bushfires. I knew the force of the hot north wind, the speed of the fire and the time needed to prepare for the actual flames. I knew how to burn firebreaks to retard the fire. With climate change, all of these factors have changed. On Black Saturday, all the known patterns of a bushfire were transcended.

Instead of a cluster of eucalyptus trees engulfed in flames, imagine a tsunami, a wall of fire crashing through towns and leaving nothing in its wake. The intensity of the typical bushfire had changed.

Instead of plumes of swirling smoke and burning leaves flying into the sky, imagine a tornado with massive balls of fire leaping over an entire valley and landing on houses on the opposite hillside. The height and force of the typical bushfire had changed.

Instead of ferocious flames fanned by a hot north wind, imagine a hurricane like Katrina, with temperatures of 110° F, blasts of over 100 miles an hour and fierce fires, like open mouths, consuming all in their path. The heat of the typical bushfire had changed.

With climate change have come increased hot spells, decreased rainfalls and unfriendly weather patterns. The rise of CO_2 in the atmosphere has led to increased vegetation in the region, much of which is tinder dry on a day like Black Saturday. We are no longer prepared for disasters like these. One hundred and seventy people were burned alive. More than 7,000 people are homeless. Graham Mills, from the Centre for Australian Weather and Climate Research is quoted in *The Australian* as saying, "the conditions that lead to extreme fire weather are heat, low humidity, wind and drought ... On Saturday the temperature set new records. ... When you get those conditions, nobody has really had experience of them ever before."[7]

---

[7] *The Australian*, 10 February 2009, at **www.theaustralian.news.com.au/story/0,,25031133-28737,00.html**, accessed June 2009.

With Black Saturday the paradigm of a typical bushfire has changed, the furor intensified and the classic bushfire scenario superseded. The "way" of the bushfire we once knew has been superseded.

This example can be multiplied. The way of the seemingly eternal ice caps has changed and seas are rising. The ways of the storm, the drought and the floods have changed. The way of the ocean is changing as villages, like those on the shores of Orissa on the Bay of Bengal, are inundated by incoming waters.

With these changes in the codes of our climate, how do we interpret our cosmos? If God were to search for wisdom, the governing design behind our climate, what would God find today? The same wisdom? A damaged blueprint? An emerging new design?

Conclusion

Assuming the above wisdom cosmology is relevant in an age when we speak of genetic codes, habitats and interdependent ecosystems, what would wisdom mean, both in terms of human searching and divine design? Or, in terms of the narrator of Job, where can wisdom be found? What would wisdom mean as we face the colliding codes in a changing climate? How do we redress the balance? Or, if wisdom herself were to speak, would she say again, "I am the way that precedes the works, the presence that informs the process?"

James Lovelock may respond by claiming that earth herself will cry aloud and take revenge on humanity. He claims that "global heating" is pushing us to the brink of destruction. "We live in a planet," he claims," that can respond to the changes we make either by canceling the changes or canceling us."[8] The issue before us in this context is how wisdom will respond and just as importantly, how will we respond.

Perhaps the answer is that if we wish to acquire wisdom in the current climate crisis we need to review each of the statements summarizing the wisdom cosmology given above, and ask whether we respect this truth and how we might to restore harmony with each code—be it genetic or other—with each domain or ecosystem and with that mysterious integrating design or blueprint that we call wisdom.

---

[8] James Lovelock, *The Revenge of Gaia* (London: Allen Lane, 2006).

Black Saturday[1]

Norman Habel

Amazing flames that scorch the sky,
Like hurricanes of fire,
Alive with eucalyptus oil,
Are roaring higher and higher.

These swirling balls of oil ablaze,
That leap o'er trees at will,
Descend on fields and flock and homes,
Explode and burn and kill.

Where's God in all this swirling ash?
Where's God in all this pain?
Awaiting somewhere in the sky
To one day send some rain?

The face of God is burnt and black;
The hands of God are red!
The God we know in Jesus Christ
Is bleeding with the dead.

Is this, O God, the shock we need
To face our life ahead,
Adjusting to a Greenhouse Age
When we must share our bread?

Christ, show us now your hands and feet,
The burns across your side
To show you suffer with the Earth,
By fires crucified!

---
[1] Tune: *Amazing Grace*.

God's Lament for the Earth: Climate Change, Apocalypse and the Urgent Kairos Moment

Barbara Rossing

> As we face rising waters, hunger, and displacement,
> God suffers with us.
> As we mourn the distress and wounds of God's creation,
> God weeps with us.
> As we struggle for justice,
> God struggles with us.
> As we expose and challenge climate injustice,
> God empowers us.[1]

A recurring question among victims of climate change in Asia, the Arctic and other regions is, Why is God punishing us?

In the Puri district of India, the village of Chhenu has already been submerged by rising seas and storm surges. Other villages, such as Udaykani, Rahakandal and Bali Bonfalo, face future relocation because of climate change. Village women took our hands to show us the encroaching ocean, just beyond the edge of their villages. Rows of little casuarina trees, planted by the local Women's Forest Protection Committee with help from Lutheran World Service (LWS), hold back the erosion of the shoreline dunes.[2] One village's well, located next to the Hindu temple, became undrinkable two years ago due to salination from a rising water table. Now there is only one private well for drinking water.

---

[1] *Call to Worship*, Eucharist Worship Service, 19 April 2009, Puri, India, written by George Zachariah and recent graduates of Gurukul Lutheran Theological College.

[2] The connections between gender and climate change are striking. In Tanzania, "red-eyed women" have been accused of being witches. Women and girls who have to walk longer distances for water because of deforestation and draught are more vulnerable to assault and rape. See, for example, *Resource Guide on Gender and Climate Change* (United Nations Development Programme, 2009) at **www.ungei.org/resources/files/genderandclimate.pdf**, accessed June 2009.

Whether in Puri, India, or in the Arctic Lutheran village of Shishmare thousands of miles away, severe storm surges and rising seas are forcing entire communities of people, with whom the Lutheran communion relates, to become climate refugees. Houses topple into the sea, traditional fishing and agricultural economies are decimated, and villagers are forced to relocate. In Tanzania, mosquitoes have arrived in villages on the slope of Mt Kilimanjaro that had hitherto never experienced malaria. Medical effects such as the increasing spread of malaria and other vector-borne diseases such as dengue fever have led medical organizations to name the climate crisis as the most urgent public health crisis facing our world.[3]

Traditional village communities throughout the world are experiencing the cruelest injustice of global climate change: it is the world's poorest people—those who have done the least to cause the problem of climate change—who are the first to suffer its catastrophic effects.

Where is God in this crisis? "I think nature is paying us for our sins," one village leader in Orissa reflects.[4] Deep spiritual questions about God's disfavor and punishment are a common response to the experience of calamitous disruptions in normal weather patterns, whether among Christian, Hindu, Buddhist, Muslim or indigenous villagers.

Participants in the LWF encounter in Puri, Orissa, India, prayed a litany that brought forward a different theological perspective—that of lament—in response to the rising waters and displacement caused by climate change. This litany gives voice to a crucial theological and biblical insight: that "God weeps with us. God suffers with us. God struggles with us." God laments with us…. But God does not curse us.

To be sure, it is possible to read the Bible in a way that interprets catastrophic weather as God's curse or punishment for people's sin. Some verses from the Book of Revelation and other biblical texts can be read to foster such an interpretation. But especially since the people most severely affected are not the same people whose sin caused the problem, interpretations of climate change in terms of punishment for sin can be highly problematic. In seeking a more pastoral biblical response we

---

[3] The Lancet Commissions in Great Britain recently concluded that, "Climate change is the biggest global health threat of the 21st century." See, The Lancet and University College London Institute for Global Health Commission, *Managing the Health Effects of Climate Change* (16 May 2009), p. 1693, at **www.ucl.ac.uk/global-health/ucl-lancet-climate-change.pdf**, accessed June 2009.

[4] Richard Mahapatra, "Climate Change in Orissa—Part Two," at **www.worldproutassembly.org/archives/2006/04/climate_change_5.html**, accessed June 2009.

can draw on other scriptural traditions, exploring how God is lamenting with us, crying out on behalf of the earth and its communities against imperial oppressors.

In this article, I shall draw on the apocalyptic and prophetic voices of Scripture to develop that interpretation that God laments with us and with the earth, crying out against oppression. I will primarily focus on the Book of Revelation. I will also draw on the prophet Jonah and the story of the Ninevites' repentance as a positive model for systemic change on the part of a major world empire facing imminent destruction, in response to God's urgent prophetic word.

In seeking to read the story of Revelation in a way that foregrounds God's urgent grief and lament for our world rather than God's punishment or curse, I will emphasize four elements of the book's message, focusing on the final depiction of the New Jerusalem as the healing vision toward which the entire book builds:

- First, in Revelation, there are frequent statements of "woe" over the earth. The intent of these statements is not to pronounce God's curse against the earth. Rather, these declarations are a divine lament or cry on behalf of the world—bemoaning the devastating conquest of earth by the unjust Roman Empire. The Greek word *ouai* therefore is best translated not as a pronouncement of "woe" but rather as a cry of mourning, "How awful!" or "Alas!"

- Second, a strong sense of an impending "end" pervades the entire Book of Revelation; however, the "end" that the book envisions is not primarily the destruction of the earth or the end of the created world. Rather, Revelation envisions an end not to the earth itself but to the imperial order of oppression and destruction.

- Third, the plagues of ecological destruction in the Book of Revelation are modeled on the plagues of the story in Exodus of Israel's liberation from oppression in Egypt, with the first-century Roman Empire cast in the role of Egypt. As with the Exodus story, the plagues in Revelation are warnings to repent, not predictions of devastation for its own sake. Their goal is liberation, not environmental destruction. As in the story of the prophet Jonah's preaching to Nineveh, if the oppressors repent, the terrible plagues will not be carried out.

- Fourth, Revelation culminates in a final vision of the New Jerusalem, one of the most earth-centered visions of our future in the whole Bible—an apocalyptic wake-up call that can renew our sense of ethical urgency and deepen our ecological commitments. Revelation places the Christian community at an ethical crossroads—a *kairos* moment—facing a choice between Babylon/Rome or citizenship in God's New Jerusalem. The book's urgent call to renounce empire and participate in God's healing and renewal gives a model for responding to the climate change crisis today.

"Alas" for earth, not "woe" upon earth: God does not curse the earth

A first step toward a more ecological reading of Revelation involves a reconsideration of the so-called "woes." If God cares about the earth and its inhabitants, then how are we to explain the apparent "woes" against the earth that are so prominent in Revelation?

The Greek word that is usually translated "woe" (*ouai*) is frequent in Revelation, beginning with the fourth trumpet in the middle of terrifying Exodus-like plagues (Rev 8:13). The word is invoked both as an exclamation and, somewhat peculiarly, as a noun, bearing the definite article: "the first *ouai*" (Rev 9:12) and "the second *ouai*" (Rev 11:14). The terrifying exclamations of "woe" throughout Revelation's middle chapters have led some interpreters to think that God has consigned the earth to suffer plagues of ecological disaster and destruction. For example, between the fourth and fifth trumpets an eagle flying through mid-heaven cries out "Woe, woe, woe to the inhabitants of the earth" (Rev 8:13). Later, a heavenly voice announces what sounds like a curse: "But woe to the earth and the sea, for the devil has come down to you with great wrath, because he knows that his time is short!" (Rev 12:12).

However, in these so-called "woes" of Revelation, God is not pronouncing a curse but rather offering a lament, bemoaning earth's suffering and abuse. In my view, Revelation's "woes" must be read in light of the book's overall critique of empire.

The Greek word *ouai* is not easy to translate into English. It is a cry or sound in Greek that can be used to express lament or pain—like a mourner keening in grief, wailing out repeated cries of "oh, oh, oh" at the death of a loved one.[5] Spanish Bibles simply use the sound "Ay, ay, ay." In my view, the Greek word *ouai* is better translated consistently as "alas!" or "how awful!" rather than "woe" throughout the entire Book of Revelation. Lamentation or "alas" is clearly the sense of the word *ouai* that is used later in chapter 18 in the three-fold formulaic lamentations pronounced by the rulers, merchants and mariners weeping over Babylon. For example, the kings of the earth say, "Alas, alas, the great city, Babylon, the mighty city! For in one hour your judgment has come" (Rev 18:10), a lament echoed by the merchants and mariners (Rev 18:16, 19).

Most translators render the three groups' expression as "alas, alas" (18:10, 16, 19, *The Revised Standard Version* and *The New Revised Standard Version*).[6] This standard translation of *ouai* as "alas" in Revelation 18 should inform our translation of other references to *ouai* in Revelation as well. The so-called "woes" then declare not a curse against the earth, but rather God's lament on behalf of the earth that has been subjugated by evil powers. It is as if God were crying "ouch" or "alas" on behalf of our suffering world: "Alas, earth and sea, for the devil has come down to you in great wrath, because he knows that his time is short!" (Rev 12:12).

Although no English word is the exact equivalent of the Greek, the subtle but important distinction between a pronouncement of "woe" and a lament of "alas" makes an enormous difference both ecologically and spiritually. "Alas" conveys a level of sympathy and concern for the earth that the English word "woe" does not. Moreover, there is no "to" in the Greek text, so typical translations of "woe to the earth" are particularly inaccurate.[7]

If we translate *ouai* as "alas," God can be understood as sympathizing in mourning and lament over the earth's pain, even while God is threatening plagues as a means to bring about the earth's liberation from injustice. Such a translation is supported by recent interpretations of similar passages in the Old Testament, such as Jeremiah 12:7–13, about

---

[5] So Margaret Alexiou, *The Ritual Lament in Greek Tradition* (Cambridge: Cambridge University Press, 1974).

[6] The Jerusalem Bible translates the laments of Revelation 18 as "Mourn, mourn, mourn," a helpful translation for underscoring the tone of lament—although the verb is not an imperative.

[7] Grammatically, in the Greek of Rev 8:13 and 12:12 the word "earth" is in the accusative case, not the dative case that would normally translated as "to." The accusative is mostly likely an accusative of reference. In light of this, a literal translation might be, "Alas, with respect to the earth" or "Alas for earth."

which Terence Fretheim has written: "these verses are a divine lament, not an announcement of judgment."[8]

End of empire, not end of earth: liberation of earth from captivity to Rome

So why does God lament or mourn on behalf of the earth in Revelation? In my view, the cries of "alas" in 12:12 and throughout the middle chapters are best understood as part of the book's larger political critique against Roman imperialism. This leads to my second point: Revelation's primary polemic is not against the earth as such, but against the exploitation of the earth and its peoples. The voice from heaven expresses God's cosmic cry of lamentation because God is outraged that the lands and the seas have been subjugated by Satan's emissary, the Roman Empire. God cries out in a cosmic lament against the violent conquests and predatory economic system of the empire that has enslaved both people and nature.

Crucial to such an anti-imperial reading of Revelation is God's proclamation that the time (*kairos*) has come "for destroying those who destroy the earth" (Rev 11:18). This statement attributes blame for the destruction of earth not to God but to unjust "destroyers" who decimate and devastate the earth. What God plans to destroy, according to this crucial verse, is not the earth itself but rather the idolatrous "destroyers" of earth—that is, Rome, with its entire political economy of exploitation and domination. This makes a crucial difference both eschatologically and ecologically for the way in which we interpret the book.

In the view of Revelation, God will no longer tolerate Rome's destruction of the earth, despite Rome's claim to rule forever. In fact, the author's so-called "end times" language was probably chosen deliberately in order to counter Rome's imperial and eschatological claims to eternal greatness. Rome claimed eternal dominance over the entire world, with such slogans as *Roma Aeterna*—eternal Rome. The boast of the whore of Babylon/Rome reflects this imperial hubris, "I rule as a queen; I am no widow, and I will never see grief" (Rev 18:7). This arrogant boast sets up Babylon/Rome for its catastrophic dethronement

---

[8] Terence Fretheim, "The Earth Story in Jeremiah 12," in Norman Habel (ed.), *Readings from the Perspective of Earth*, The Earth Bible 1 (Sheffield: Sheffield Academic Press, 2000), pp. 96–110.

and destruction only a few verses later. God answers Rome's boasts of omnipotence and eternity with a resounding "no." In response to the question of the eternity of Roman rule, "Sovereign Lord, holy and true, how long will it be ... ?" (Rev 6:10), Revelation comforts the souls who had been martyred by Rome with the message that it will be "a little longer" (Rev 6:11) until God will destroy "those who destroy the earth" (Rev 11:18). Revelation's insistence on the imminent "end" assures its audience that Rome will not rule the earth forever. God's *kairos* moment puts an end to oppression.

Revelation's lament, its "Alas for the earth" (cf. Rev 12:12), concedes that Rome's own imperial claims of domination over the earth have come to pass. But Revelation makes clear that Rome's unjust exploitation of the earth is only temporary. Satanic Rome will not last forever. The devil knows that "his time is short" (Rev 12:12). The cry of "alas" (*ouai*) "for the earth" (Rev 12:12) expresses the certain hope that Satan/Rome will not stalk the earth much longer.

In summary, the God of Revelation does not seek to destroy the earth. Rather, God seeks to rescue the earth—the land, the seas and the creatures who inhabit them—from the sickness of empire that is devastating the world, so that creation can be brought to fulfillment.

The Exodus story in Revelation: plagues as warnings

How will the liberation and healing come about for the earth and its peoples? The most important biblical model for Revelation is the Book of Exodus, the story of the liberation of Israel from bondage in Egypt. Imagery from the biblical Book of Exodus furnishes the pattern for much of Revelation's imagery, including Jesus as the Lamb who takes on the role of Moses.[9] The connection to Moses and the Exodus becomes explicit when God's servants sing the "song of Moses the servant of God and the song of the Lamb" (Rev 15:3). The entire Book of Revelation suggests a parallel between the Christians' journey out of Rome and the Israelites' journey out of Egypt. The author of Revelation calls Christians to "come out" of Babylon (Rev 18:4). As such, the Book of Revelation

---

[9] For the argument that Revelation draws most extensively on Exodus traditions, see the works of Elisabeth Schüssler Fiorenza, especially *Revelation: Vision of a Just World* (Minneapolis: Fortress Press, 1991).

gives a "rereading of the Exodus, now being experienced not in Egypt but in the heart of the Roman Empire."[10]

In Revelation's rereading of the Exodus story, the Roman Empire is scripted in the role of the predatory system analogous to ancient Egypt. As Ellen Davis points out, biblical writers aptly called Egypt "the iron furnace" (cf. Deut 4:20; Jer 11:4; or "iron-smelting furnace," *New International Version*). Egypt was "the biblical archetype of the industrial society: burning, ceaseless in its demand for slave labor (the cheapest fuel of the ancient industrial machines), consuming until it is itself consumed."[11] The exodus liberated God's people and healed them from the sicknesses of Egypt (cf. Ex 15:26). John of Patmos applied that biblical critique of Egypt as a "sick society" to diagnose as a sickness the all-consuming Roman imperial economic system of his day. Promises of "manna" (Rev 2:17) and the tree of life for the "healing of the nations" (Rev 22:2) served to encourage the Christian community in their exodus out of the Roman system, on their journey towards healing.

In our own time, we can ask what manifestations of empire analogous to Egypt or Rome may pose comparable threats to the health of the world and to God's people today. From the perspective of climate change, for example, we might diagnose as "Egypt" our unsustainable system of carbon consumption that poisons and enslaves the world, imposing a kind of climate slavery on the world's poorest nations and on future generations. The healing vision of New Jerusalem (Rev 21–22) can help us envision an alternative way of life in contrast to the "iron-smelting furnace" of consumption.

Understanding the profound ways that Revelation borrows from the Exodus story can also help us to interpret what is perhaps the most ecologically difficult imagery of the book—the plague sequences described in the middle chapters (Rev 6–16). As we saw with regard to the "woes," Revelation's terrible plagues can give the impression that the destruction of rivers, scorching heat, burning of forests, waters turning to blood and other environmental calamities are somehow an expression of God's will to destroy the earth. But Revelation's plagues are threats and warnings to oppressors, not predictions of inevitable destruction. They are modeled on God's threats of punishment against

---

[10] So Pablo Richard, *Apocalypse: A People's Commentary on the Book of Revelation* (Maryknoll: Orbis, 1995), p. 77.

[11] Ellen Davis, *Scripture, Culture, and Agriculture: An Agrarian Reading of the Bible* (Oxford University Press, 2009), p. 69. For the description of "the sick society that is Egypt," see p. 71.

Pharaoh in Exodus. The plagues serve as wake-up calls, warning of the consequences of Rome's unjust actions. God does not predict that these ecological disasters must happen—they are rather urgent warnings of what may happen if oppressors do not repent. The plagues project into the future the logical consequences of the trajectory that the Roman Empire is on, so that people can see in advance where the dangerous imperial path is taking them. The plague visions of Revelation function like the nightmarish visions Ebeneezer Scrooge experiences in Charles Dickens's *A Christmas Carol*—they show a terrifying future that will happen if Scrooge does not change his life. But they also make clear that there is still time for change, and that disaster is not inevitable.

The plagues of Revelation are part of the book's liberating vision—they are "ecological signs" enlisting nature itself in the struggle for liberation. Terence Fretheim argues that the Exodus plagues "function in a way not unlike certain ecological events in contemporary society, portents of unmitigated historical disaster."[12] In our time, it can be especially important to see the threats of rising seas, drought and other calamities as wake-up calls—portents of disaster to warn of the consequences of climate change, and to call upon industrial nations to change course before it is too late.

Chilean scholar, Pablo Richard, interprets Revelation's plagues as imperial assaults on the poor, arguing that it is inaccurate even to call the plagues of Revelation "natural" disasters:

> In earthquakes and hurricanes the poor lose their flimsy houses because they are poor and cannot build better ones; plagues, such as cholera and tuberculosis, fall primarily on the poor who are malnourished. ... Hence the plagues of the trumpets and bowls in Revelation refer not to "natural" disasters, but to the agonies of history that the empire itself causes.[13]

Richard draws an analogy also to contemporary imperial situations, compiling a list that could well be updated to include the climate crisis: "Today the plagues of Revelation are rather the disastrous results

---

[12] Terence Fretheim, "The Plagues as Ecological Signs of Historical Disaster," in *Journal of Biblical Literature* 110 (1991), p. 387.

[13] Pablo Richard, *Apocalypse: A People's Commentary on the Book of Revelation* (Maryknoll: Orbis, 1998), p. 86.

of ecological destruction, the arms race, irrational consumerism, the idolatrous logic of the market."[14]

Threats of dire consequences await oppressors if they continue in their unjust ways. As Revelation 16:5–6 shows, it is axiomatic or fitting (*axios estin*) that the consequences of oppression on the earth will come back around in boomerang-like fashion upon those who commit injustices. We can apply a similar logic of consequences to describe the danger of global climate change. Global warming is not punishment from God, but rather a consequence of the physical fact that in this universe that God has created, with its finely-tuned atmosphere, certain actions cause other things to happen. It is a physical fact that carbon dioxide, methane and other greenhouse gases trap heat. In terms of a biblical logic of consequences, what may be "axiomatic" (*axios estin*, Rev 16:6) for our world today is that we cannot continue on this trajectory of carbon consumption without heating up the planet to dangerous levels. We must alter the course of our life before it is too late.

To summarize, I am arguing that both Revelation's plagues and the "woes"—two elements of the book that can seem the most anti-ecological—show us God's cry for a world that needs to be freed from the toxic system of imperial exploitation. As in the Exodus story, God calls on people to "come out" of empire, to withdraw from participation so as not to be implicated in empire's sins (Rev 18:4) and so as to be able to participate in the healing vision of renewal—God's New Jerusalem.

The present moment as a kairos time: the call for repentance and testimony

Revelation's focus on the urgency of the present moment is another aspect of the book that can help us face the crisis of global climate change. The Ecumenical Patriarch Bartholomew I, Archbishop of Constantinople, has called the climate change crisis a "*kairos* moment" for churches and for the world. He warns that the time is short for the world to take decisive action on climate change:

> As individuals we are often conscious of a *kairos*, a moment when we make a choice that will affect our whole lives. For the human race as

[14] *Ibid.*

a whole, there is now a *kairos*, a decisive time in our relationship with God's creation. We will either act in time to protect life on earth from the worst consequences of human folly, or we will fail to act.[15]

Archbishop Bartholomew ended the 2007 Symposium on the Arctic in Greenland with this sobering prayer: "May God grant us the wisdom to act in time."

Dr Martin Luther King Jr used the expression "the fierce urgency of now" to refer to the US civil rights movement of the 1960s—an expression that could also be used to frame the climate crisis as a civil rights crisis today.[16] That is because scientists tell us that we are soon reaching thresholds of carbon dioxide levels past which it will be impossible to reverse, runaway catastrophic changes, such as the melting of the Greenland and West Antarctic ice sheets, or the melting of the methane rich Arctic permafrost. Consequences of climate change will fall most heavily on the poorest people of the world. The 2007 Intergovernmental Panel on Climate Change (IPCC) report recommended the safe level of atmospheric carbon dioxide at no more than 445 parts per million. A growing number of scientists now recommend an even lower target for atmospheric carbon dioxide and other greenhouse gases, 350 parts per million, because of worse-than-expected changes since 2007.[17] The world's premier climate scientist, James Hansen, believes there is still time to avert irreversible tipping points. But in Hansen's view we have less than ten years to act.[18]

---

[15] His All Holiness Ecumenical Patriarch Bartholomew, Symposium on the Arctic, 7–12 September 2007, Greenland: "The Mirror of Life: Part 3, Symposium Closing Address," at **http://orth-transfiguration.org/library/orthodoxy/mirros**, accessed June 2009.

[16] Dr Martin Luther King Jr, "A Time to Break Silence," April 4, 1967, in James M. Washington (ed.), *A Testament of Hope: The Essential Writings and Speeches of Martin Luther King, Jr.* (San Francisco: Harper San Francisco, 1986), p. 243: "We are confronted with the fierce urgency of now. In this unfolding conundrum of life and history there is such a thing as being too late. Procrastination is still the thief of time. ...Over the bleached bones and jumbled residue of numerous civilizations are written the pathetic words: 'Too late.'"

[17] See the Web site of the 350 organization, at **www.350.org/**. The 350 parts per million target is based on climate scientist James Hansen's 2008 argument in, "Target atmospheric CO_2: Where should humanity aim? *Open Atmos. Sci. J.*, 2 (2008), pp. 217–31. See also Bill McKibben, "Earth at 350," in *The Nation*, 12 (May 2008).

[18] James Hansen, "Why We Can't Wait," in *The Nation*, 7 May 2007. Some organizations now speak of an even shorter timeframe of 100 months within which the world must stabilize and begin significantly to decrease carbon emissions in order to avoid tipping points. See the Web site **http://onehundredmonths.org/** whose partners include a number of Christian organizations.

The church must take seriously such mounting evidence from science. The church must name this ten-year window as a *kairos* moment, a moment of hope and urgency.

How can the church draw on the Bible publicly to address this crisis with hope, underscoring especially the urgency of that ten-year window? The biblical Book of Revelation holds out hope for the future, both with its New Jerusalem vision and it repeated calls for repentance. Revelation is a hopeful book in the sense that it believes that there is still time for people to "come out" of empire (Rev 18:4) and live according to God's vision for the world. It is not too late for repentance. This explains why Revelation departs from the Exodus tradition at one crucial point. While Revelation largely follows the Exodus story for its terrifying plagues—the seven-fold sequences of trumpets and bowls that are poured out upon the earth and its waters in chapters 8–9 and 15–16—Revelation refrains from using the exodus motif of the hardening of Pharaoh's heart. Hearts are never hardened in Revelation.

To be sure, the book's positive calls for repentance (the imperative of *metanoe*, to repent, Rev 2:5, 16, 3:3, 19) are concentrated in the seven opening letters, whereas later references to repentance are all phrased in the negative ("they did not repent of ..." Rev 9:20, 21; 16:9, 11). Yet, Elisabeth Schüssler Fiorenza has made a persuasive case that even these negative references to repentance in chapters 9 and 16 serve as part of the book's rhetorical appeal to the audience to repent: "John writes this grotesque and brutal vision not for cruelty's sake but rather for the sake of exhortation to repentance."[19]

In chapter 11, Revelation lifts up a model of successful repentance in the rest of the people—nine tenths of the population—who do respond to the testimony of the two witnesses and are persuaded to give glory to God (Rev 11:13). This turning on the part of the populace is a hopeful aspect of the book from which we can draw. The two witnesses of Revelation symbolically represent God's people. These two witnesses give the kind of public testimony and witness that John wants the Christian community of his own time to emulate in their own prophetic testimony against the worship of Rome. Such public testimony can furnish a model for the church today. As African American scholar Brian Blount describes,

---

[19] Schüssler Fiorenza, *op. cit.* (note 9), p. 72.

"John is calling for a witness of active, nonviolent resistance to Rome's claim of lordship over human history."[20]

Perhaps an analogy in today's times would be a call for the church to engage in a witness of active nonviolent resistance to our "worship" of fossil fuels, our addiction to an unsustainable carbon based economy. The church must play the role of the two witnesses today, calling for such a public *metanoia* or repentance. Scientists tell us that halting carbon emissions at a safe level is fully possible with existing technology.[21] What is needed is public, political commitment to reducing carbon emissions by eighty to ninety percent below 1990 levels by the year 2050. The Council of the Lutheran World Federation has called for reducing emissions by forty percent by the year 2020, a goal that is both hard-hitting and achievable—if we muster the political will.[22] The Uppsala Interfaith Climate Manifesto 2008, convened by the Church of Sweden, calls for reductions of forty percent below 1990 levels by 2020 and ninety percent reductions by 2050.[23] Other churches and religious communities have issued similar statements. We must all join together as religious communities, scientists, public policy makers and other prophetic leaders to give public testimony that calls upon the world—and especially its richest nations—to turn away from our addiction to a dangerous, carbon-consuming way of life.

Time is of the essence in Revelation. But interestingly, the book's perspective is not simply of time hurtling towards an inevitable end. Rather, Revelation puts great emphasis on the importance of the present moment as a moment for decision and repentance, a fact noted by a number of scholars. Shifts from past tense to present and future, along with calls for repentance and use of deliberative rhetoric, all serve to draw the audience into what Canadian Lutheran scholar, Harry Meier, calls "an abiding sense of the imminent"—extending the urgency of the

---

[20] Brian K. Blount, *Can I Get a Witness? Reading Revelation through African American Culture* (Louisville: Westminster John Knox, 2005), p. 40.

[21] *IPCC Working Group III Summary Report for Policymakers*: "The range of stabilization levels assessed can be achieved by deployment of a portfolio of technologies that are currently available and those that are expected to be commercialized in coming decades," at **www.ipcc.ch/ipccreports/ar4-wg3.htm**, accessed June 2009.

[22] Resolution adopted at the 2008 meeting of the LWF Council, Arusha, Tanzania.

[23] *Hope for the Future: Uppsala Interfaith Climate Manifesto 2008: Faith Traditions Addressing Global Warming* (Church of Sweden, 2008), at **www.svenskakyrkan.se/Webbplats/System/Filer/14AD960A-3652-460F-8F9E-E31BED75158D.pdf**

present moment.[24] The entire Book of Revelation calls on the audience to "come out" of empire before it is too late (Rev 18:4), in order not to fall prey to the catastrophic judgment and plagues, in order not to share in the collapse of the empire.

Are there biblical models for such a drastic turning, analogous to what the world's citizens must undertake in the next ten years? The most intriguing biblical model may be the empire of Nineveh in the Book of Jonah. Unlike Revelation's view that the Roman Empire itself is hopelessly doomed, Jonah offers a more positive storyline for imperial repentance and turning. The huge imperial capital city of Nineveh changed its course and averted certain disaster in just forty days. Today, in the face of naysayers who say we can do nothing, or who claim that climate change is not dangerous, Nineveh can offer us hope. Nineveh can serve as a parable for how the greatest "empire" on earth today can shift course and avoid disaster. Jonah's urgent warning to Nineveh that it had "just forty days" recalls the warnings from our best scientists that we have "less than ten years" to avert irreversible climate tipping points such as the melting of the Greenland and West Antarctic Ice Sheet.

Nineveh's transformation began at the grassroots level. It was the regular people in the streets who first believed God and responded to the prophet's warning, engaging in public action (Jn 3:5). Their grassroots action then caught the attention of the king who repented and issued policy directives for public repentance. Humbly, the king said, "this is what we have to do." He embarked on a fast-track campaign to change public will. The king made his case: "Perhaps it will be in time. Hopefully we are in time to avert the disaster so we will not be destroyed." The Book of Jonah shows how thanks to grassroots mobilization, good leadership, and an effective prophet, a giant ship of state—including all the people and even the animals—can abandon its destructive course in time to avert a catastrophe. We can learn from Nineveh.

Time is short, our scientists tell us today. Scientists may be our prophets today who are preaching "just forty days," or "just ten years." The question is whether we will heed their prophetic testimony. Revelation places the Christian community at an ethical crossroads—a *kairos* moment—facing a choice between the doomed empire of Babylon/Rome or citizenship in God's New Jerusalem. The heart of the message of Revela-

---

[24] Harry Meier, *Apocalypse Recalled: The Book of Revelation After Christendom* (Minneapolis: Fortress, 2002), p. 147.

tion is God's invitation to us to come into the vision of the New Jerusalem, embracing its promise of healing for our world rather than destruction. Revelation also calls upon the Christian community to take up the role of the two witnesses of chapter 11, voicing our prophetic testimony about the dangerous trajectory that the world is currently on. Will our churches take up that urgent call to turn around our "empire" to a more sustainable path, in the same way that Nineveh repented and turned? Or, will we follow the "iron-smelting furnace" trajectory of the doomed empires of Egypt and Rome? There is still time for healing our world, scientists tell us. We have not yet crossed irreversible thresholds.

Archbishop Desmond Tutu preached a sermon at a June 2007 United Nations climate conference in Tromsø, Norway, hosted by the Church of Norway. The service included the testimony of a young woman from the Pacific island nation of Kirabati, a country that will soon be submerged due to climate change. "We hold the planet's future in our hands," said Archbishop Tutu, basing his sermon on Deuteronomy 30:19, Moses' final sermon to the people. "'I call heaven and earth to record this day against you, that I have set before you life and death, blessing and cursing: therefore choose life, so that you and your descendants may live."

For the sake of our sisters and brothers whose homes are already threatened by the effects of climate change, and for the sake of future generations who can still be saved, we must choose life. Our prayer must be the prayer of Archbishop Bartholomew: "May God grant us the wisdom to act in time."

Cross, Resurrection and the Indwelling God

Cynthia Moe-Lobeda

Climate change presents a moral crisis never before faced in the history of our young and dangerous species. We have become a threat to life on earth. Our numbers and our excessive consumption threaten the earth's capacity to regenerate life. God created a planet that spawns and supports life with a complexity and generosity beyond human ken. According to the creation stories of Genesis, God said, "it is *tob*," life furthering. Never before has one species endangered that generative capacity. We—or rather, some of us—have become the "uncreators."

Climate change: a matter of "privilege"

The suffering and death caused by climate change is not evenly distributed among the earth's human creatures. On the contrary, those dangers are structured into people's lives along the same axes of oppression that structure social injustice. People not considered "white" and those who are economically impoverished are at far more risk. Said differently, privileges of color and class offer to a few of us relative protection from the earliest and severest impacts of global climate change. While we all may be in this together, initially we are not all in it in the same way or to the same deadly extent.

This reality is gut wrenching for people of relative economic privilege, like myself, who live in the global North. Our lives are wound up in and benefit materially from social structures and norms that breed deadly ecological destruction and economic violence for many whom we fail to see. Everyday life, in the ravenously consumptive and petroleum dependent mode that we consider normal, threatens the web of life called forth by the One whom we know as Creator of all.

The world's great faith traditions are called to plumb their depths for wisdom to contribute to the great moral challenge of our day: to forge ways of being human that allow earth to flourish and all people to have the necessities for life with dignity. I believe that all religious traditions have particular gifts to bring to the table, and are called to put these in dialogue with each other and with other bodies of knowledge, including the natural and social sciences.

A theology of the cross for the "uncreators"

Christians who walk in Luther's footsteps have an invaluable contribution to make through a theology of the cross and resurrection, especially as it is linked to Luther's notion of God dwelling in all of creation. Amid widespread and vast suffering caused by global warming, in which citizens of the USA play such a disproportionate role, the cross is central if Christians of relative economic privilege in the USA, or elsewhere, are to play a significant role in constructing earth honoring ways of living.

But first, historical misuse of the cross issues a warning. There are reasons to distrust many theologies of the cross. The question is, Which cross and whose cross? An hermeneutic of trust and of suspicion need to work together. Led by our forbearers and by Jesus himself, we stand in a critical tradition, testing our claims and convictions. To what extent do they convey or betray the splendid mystery of God's unbounded and undefeatable love for this good creation and presence with and within it, or do they betray that good news?

False crosses have been with us since at least 313 AD when Christianity became the religion of the reigning imperial power of the "known" world. The cross of Constantine, for seventeen hundred years justifying war in the name of God, was not and is not the cross of the God revealed in Jesus of Nazareth. The cross of the "white Christ," known most horrifically in the American slaveholders' religion, betrayed the cross of Jesus Christ.[1] That cross is also present today in well-intentioned pictures of a North European Jesus, subtly linking whiteness with goodness and saving power. The cross of "bear your suffering meekly," "like a lamb," when it drives abused women and others back into the hands of their abusers is not the cross of Jesus. Nor is the cross of Christian religious supremacy, raised in communities where the faith of Jews, Muslims, or people of other religious traditions is denigrated.

The medieval turn to a cross with Jesus nailed to it, forever dead or dying, bears another danger: the risen Christ, who is alive and breathing

---

[1] See Kelly Brown Douglas, *The Black Christ* (Maryknoll: Orbis Books, 1994).

in and through creation, is lost. The Incarnate One revealed today in a grain of wheat, in the touch of wind or sun on bare skin and in human goodness—this Christ with and within us is pushed under when the cross holds a dead and captive Christ.[2]

Indeed, false crosses abound, in history and today. At worst, they have justified domination, exploitation and dehumanization. The responsibility of "faithful disbelief" includes recognizing and exposing these falsehoods. Yet, the cross and resurrection make the central story of Christian faith ultimately and eternally life-giving.

How might the cross, and our understandings of it, help move people of faith to work toward ecologically sustainable and socially just ways of life? How might the cross of Jesus Christ contribute to an all-encompassing transformation of society—the reformation of economic policies and practices, political structures, modes of transportation and recreation, architecture, business and more?

To pursue that question, we need to consider what currently undermines our capacity to work toward ecologically sustainable and socially equitable ways of life. What disables our moral and spiritual capacity to live as if we truly believe that God loves this good creation with a love that seeks well-being for all, and that we are called to embody that love? In short, what accounts for our astounding moral inertia in the face of earth's distress and the anguishing poverty of multitudes? These crucial questions defy simply answers, and are highly contextual. The context out of which I write is that of US Christians who are "relatively secure economically."[3]

This essay proceeds in two parts. I first respond to the above question by probing four key factors undergirding moral inertia. Subsequently, I argue that a theology of the cross could counter those four factors, thus enabling us to live in more sustainable and just ways.

---

[2] For this insight, I thank Rita Nakashima Brock and Rebecca Anne Parker, in their book, *Saving Paradise* (Boston: Beacon Press, 2006).

[3] With this phrase, I mean those of us who have some degree of choice in how we spend our time, energy and material goods, and who have relative mobility and access to democratic processes.

What contributes to moral inertia?

Multiple factors contribute to moral inertia. In a previous work, I have investigated two.[4] Others would include the power of sin in human life and the practical constraints of time. Here I suggest four additional factors:

Avoidance and denial of our participation in structural sin

Some of us are not engaged because we fail to recognize the role that we play in the ongoing sins committed against the earth and against the many people whose lands, resources, and labor make possible our lifestyles of outrageous levels of consumption. We do not see, because seeing would be too terrible and it would be too painful to acknowledge how we are implicated in profound and widespread suffering and in what threatens the life of the world today. Over 500,000 children under the age of five died in Iraq between 1991 and 1998 from disease connected to the USA's bombing (devastation of water systems and electrical system and land contamination), and US invoked sanctions prohibiting medicines from entering Iraq. How could we live with realities like this, if we were truly to take them in? How could we face the piercing anguish endured by the parents of those children? While human life depends on the health of earth's life-support systems (air, soil, water, biosphere), "every natural system on the planet is disintegrating,"[5] due in significant part to massive consumption of petroleum products in the last fifty years. We, citizens of the USA, with our blind and insatiable addiction to oil, spew forth over one hundred times the deadly greenhouse gases per capita, as do our counterparts in some other lands. How will we face our children, when they realize what we have done? How can we think the unthinkable, acknowledge what is utterly unacceptable?

Not by intent or will, but by virtue of the social structures that shape our lives, we are complicit in both eco-cide and economic brutality. It has been said that to be human is to suffer from knowing that we cause suffering. Knowing that we cause suffering is not new; knowing that we cause this magnitude of suffering is unprecedented. Our forbear-

---

[4] Cynthia Moe-Lobeda, *Healing a Broken World: Globalization and God* (Minneapolis: Fortress, 2002), pp. 30–69.

[5] Paul Hawken, *The Ecology of Commerce: A Declaration of Sustainability* (New York: Harper Business. 1993), p. 22.

ers have not prepared us for this self-knowledge of ourselves as the "uncreators."

From this kind of knowledge, we flee. Not seeing—moral oblivion—is far more bearable. Where it fails, numbness may set in, and when numbness thaws, despair lurks. We retreat into denial and defensiveness, privatized morality, or exhaustion. Holy outrage and lament are dead before they are born, and we hide from accountability for systemic sin under the comforting cloak of private virtues. Grave moral danger accompanies this avoidance and denial. When good and compassionate people do not see the consequences of their ways of life, uncritically accepting them as normal, natural, inevitable, or divinely ordained, they simply carry on with them.

Denying who we are as bearers of God's love

Others who dare to face their participation in structural sin may escape from resisting it by denying who we human beings are and why we are created: we are friends of God's, called and empowered by God to receive God's love, and live out that justice making love in the world. Life is breathed into us for a purpose. We are given a lifework: to receive God's love; to love God with heart, mind, soul and strength and to love neighbour as self. We are here to let God work through us, in us and among us to bring healing from all forms of sin that would thwart God's gift of abundant life for all. This is our vocation as Christ's body on earth today. If the first factor is a failure to see the consequences of social structural sin in our lives, this second is a failure to see the depth and extent of the freedom for which God has set us free from sin to serve the God of life.

A sense of powerlessness

For many people, moral inertia in the face of the earth crisis stems from a sense of powerlessness. At some level, many of us sense that something is terribly wrong: life should not provide unlimited consumption to some while others starve. Yet the systemic forces undergirding this and the earth's degradation seem too powerful for human agency. The sense of "not being able to make a difference" easily overwhelms. It seems impossible or at least difficult to trust that God indeed is luring all of creation toward the reign of God, and that no form of sin can ultimately triumph.

An anthropocentric lens

Finally, the anthropocentric lens through which we tend to view God's indwelling presence may inhibit the moral power inherent in it. Until recent eco-feminist theologies, feminist theologies of mutual relations, and ecological theologies, Western Protestant theology and ethics have not taken seriously the ancient Christian claim that God dwells not only within human creatures, but within "all things." Failing to consider the presence and power of God abiding in "other-than-human" parts of creation, we fail to consider how that presence might nurture the human capacity to serve God's work on earth. Furthermore, anthropocentric assumptions preclude questioning the implications that ecological destruction might have for a faith tradition that locates Christ on the underside of power and in places of destruction and pain. We do not see a cruciform earth if the crucified Christ is imaged only in terms of what is human.

To confess Christ is to profess "that little point [of the truth of God] that the world and the devil are at that moment attacking."[6] Today, those points include the truths that this earth is infinitely beloved by God, and that we are called to embody God's justice making love in all that we do. I am persuaded that the cross of Christ and living out a theology of the cross may counter the above debilitating dynamics, and thus unleash moral/spiritual power to strive for justice making, earth honoring ways of being human.

How the cross enables moral power

How might the cross counter these disabling dynamics and enable God's people to: (1) recognize the extent of our implication in ecological and economic injustice; (2) claim our identity as participants in God's life-giving and lifesaving work on earth; (3) embody a sense of hope and power for that work; and (4) receive the moral power, motivation

---

[6] Martin Luther, cited by Douglas John Hall, *Confessing the Faith: Christian Theology in a North American Context* (Minneapolis: Fortress Press, 1996), p. vi.

and wisdom that flows from the presence of God inhabiting "even the tiniest leaf?"[7]

For insight, we turn initially to the lived theology of the cross emerging in Dietrich Bonhoeffer's last two works. We then consider the ancient theology of Christ immanent in all things, as this converges with a theology of the cross. A significant assumption about theological method surfaces in Bonhoeffer's last two works, *Letters and Papers from Prison* and *Ethics*.[8] Reflecting Luther, Bonhoeffer demonstrates that theology develops out of the struggle to live the gospel wherever and whenever it is betrayed. He was convinced that moral power dies when good people fail to recognize evil in the guise of good. He warns that the moral sensibility of good people is easily warped by their failure to recognize social evil "disguised as light, charity, historical necessity, or social justice."[9] Referring to "fools" who have become passive and complicit in the face of structural evil, he writes, "The fool will be capable of any evil and at the same time incapable of seeing that it is evil. Here lies the danger."[10] Theology, then, must address the very points at which evil parading as good is attacking the life of the world, disclose evil for what it is and counter it with prayer and righteous action. As Bonhoeffer realized, to do so may be at risk of life, but is actually lifegiving. In this sense, his theological method instructs that to glimpse the meaning of the cross, we must seek and see where the good news of God's unquenchable and incarnate love for the world is betrayed, and there to act in accord with that love.

In *Ethics*, Bonhoeffer formulates four alternative ethical frameworks. All are theologically grounded and elaborate three claims inherent in the theology of the cross expressed by Luther and by Bonhoeffer in his later works:

---

[7] Martin Luther, "That These Words of Christ, 'This is My Body,' etc. Still Stand Firm Against the Fanatics (1527)," in Jaroslav Pelikan and Helmut T. Lehmann (eds), *Luther's Works*, vol. 37 (Philadelphia: Muhlenberg Press, 1961), p. 57. This 55-volume series is hereafter abbreviated as *LW*.

[8] Dietrich Bonhoeffer's theology, partly because it is inherently contextual and experiential, develops over time. Here we do not examine those shifts, but focus on Bonhoeffer as expressed in these two works. They extend the theo-ethical convictions, claims and constructions begun in his previous works, but also nuance, critique and develop them in substantive ways.

[9] Dietrich Bonhoeffer, *Letters and Papers from Prison*, ed. by Eberhard Bethge (New York: Collier Books, 1972), p. 4.

[10] *Ibid.*, pp. 4, 9.

The cross reveals us to be "curved in on self"[11]

First, people of relative privilege tend to arrange their lives in ways that hide from them the cruel impact of structural sin and our implication in it. Our media, zoning policies, investment procedures, commercial activities, vacation and recreation patterns, transportation routines and other life habits shelter us from those realities. I do not see the children of Mozambique, for example, who do not eat because their nation's resources go to finance the international debt, a debt structure that brings wealth from the most impoverished nations to the wealthier.

Mirroring Luther's theology of the cross, Bonhoeffer counsels that the work and ways of God are revealed most fully in Jesus Christ. Moreover, in some way beyond full human comprehension, Christ is known perhaps most deeply in places of brokenness and suffering. Thus, we will know and glimpse God most fully when we recognize God in the goodness and splendor of earthly life and allow ourselves to be present in profound solidarity where people and creation suffer most.[12] Where systemic injustice breeds suffering, solidarity means seeing that injustice and, in Bonhoeffer's terms, "putting a spoke in the wheel," to stop it.

To be deeply present "with and for" those who suffer the ravages of systemic exploitation, and to seek its undoing, is to begin to see the world upside down. We begin to see that ways of life previously assumed to be good, may not be. We begin to realize ourselves to be people so "curved in on self" that we accept the reality of poverty and ecological degradation, without asking about the political and economic structures that make this possible, and how they can be resisted. A theology of the cross opens our eyes to who we are as participants in systemic sin and thereby unleashes moral power.

By itself, this would not yet be a theology of the cross of Jesus Christ. To behold in the cross the depth of human corruption, but not our salvation from it, would be to defy the gospel. The condemnation would be too much to bear without at the same time realizing that precisely there, in human brokenness and bondage to sin, the saving, healing and liberating Christ is present. The only force that truly can heal creation

---

[11] *Se incurvatus en se* (self curved in on self) was Luther's phrase for describing the distortion of human life by sin. We become beings turned in on ourselves, serving self-interest in evident and subtle ways, above all else.

[12] Problems with the concept of solidarity and responses to it are sketched in Moe-Lobeda, *op. cit.* (note 4), pp. 118–223.

is drawn there, bringing forth healing power that we did not know we had. This enables us to see the structural brutality of which we are a part, without being destroyed by that knowledge. Canadian theologian, Douglas John Hall, says it well. The central message of the cross

> is not to reveal that our condition is one of darkness and death; it is to reveal to us the One who meets us in our darkness and death. It is a theology of the cross not because it wants to put forth this ghastly spectacle as a final statement about life in this world but because it insists that God ... meets, loves, and redeems us precisely where we are: in the valley of the shadow of death.[13]

This claim is stranger than it might seem. God's presence in the depths of our brokenness means that God is present with grace even there where we are perpetrators of tremendous violence against others. God is present even if we continue with that violence, and even if we have no awareness of God's presence and no faith that God is present. A central message of what became known as Luther's theology of the cross, and continued in Bonhoeffer, is that where God seems absent, there God is. God is hidden in God's apparent absence. God's saving power is hidden in the form of its opposite (*sub contrario suo abscondita sunt*). Nothing can "separate us from the love of God in Christ Jesus our Lord" (Rom 8:39). The power of this claim is immeasurable for those who have glimpsed even only momentarily the horror of being a wealthy Christian in a world filled with hungry people, whose hunger is connected to our wealth. This saving claim makes it possible to see that reality, rather than pretending that the economic systems creating our wealth are beneficial to all. When reality seems "distorted and sinful, and seemingly God-forsaken ... a theologian of the cross is not afraid to recognize reality for what it is."[14] In Luther's words, "A theologian of glory calls evil good and good evil. A theologian of the cross calls the thing what it actually is."[15]

---

[13] Douglas John Hall, *Lighten Our Darkness: Towards an Indigenous Theology of the Cross* (Philadelphia: Westminster, 1976), p. 149.

[14] Winston Persaud, "Luther's 'Theologia Crucis: A Theology of Radical Reversal' in Response to the Challenge of Marx's Weltanschauung," in *Dialog* (29:4), pp. 265–66.

[15] Luther, "Heidelberg Disputation (1518)," in *LW*, vol. 31, pp. 39–58.

The cross reveals us to be bearers of indomitable love

While the first claim revealed our identity as participants in structural sin, this claim reveals us as a dwelling place of the God revealed in Jesus Christ, and therefore as subjects of Christ's love.

A widely accepted theological understanding is that the baptized followers of Jesus Christ live in a paradoxical moral reality—the "already and not yet" reign of God on earth. The implications of this claim are vast. One, relevant here, is that while we are implicated in cruel forms of oppression and ecological destruction, we also are the body of Christ on earth. The living Christ and the Spirit of God abide within and among the people of God even while we are not—on this side of death—free from bondage to sin. Bonhoeffer is adamant that the love of Christ—revealed most fully in the cross—has chosen to "abide in" the church (although not only in the church). For him, as for Luther, the finite bears the infinite (*finitum capax infiniti*). The "finite" is all of creation, yet also the church.

In Bonhoeffer's terms, Christ dwelling in the church "conforms" it to "the form of Jesus Christ." God's overflowing love is incarnate as a believing community acts responsibly in the world on behalf of abundant life for all, and against what thwarts it.[16] This requires recognizing social evil, naming it, and "putting a spoke in the wheel" of earthly powers that demand disobedience to God. The power to serve others and resist social structural evil, even when doing so is very costly, is the actual love of God as Christ taking form in the community of faith.[17] Christians as objects of Christ's love become subjects of that love. Faith is both "faith in Christ" and "faith of Christ."[18] For Bonhoeffer, this is not a matter of efforts to "become like Jesus." Rather, it is a matter of

---

[16] For Bonhoeffer, conformation with the form of Christ implies refusing conformation with ways of life that betray Christ. Bonhoeffer's understanding of the cross bridges the gap between theologies of the cross that see Christ atoning for human sin, and theologies of the cross that see Christ executed by imperial power for his allegiance to the compassionate and justice making reign of God. For Bonhoeffer, the cross was both.

[17] Bonhoeffer writes: "The relation between the divine love and human love is wrongly understood if we say that the divine love [is] ... solely for the purpose of setting human love in motion. ... On the contrary ... the love with which [humans] love God and neighbor is the love of God and no other ... [T]here is no love which is free or independent from the love of God." Dietrich Bonhoeffer, *Ethics*, ed. by Eberhard Bethge and trans. by Neville Horton Smith (New York: Simon and Schuster, 1995), pp. 55–56.

[18] The New Testament Greek generally translated as "faith in Christ," in many instances, also may be translated accurately as faith "of" Christ.

the Spirit working to unite human beings with God in Christ. Conformation with the form of Christ is formation toward a freedom to live as whole persons, to be fully human, to be the Creator's creatures, and to help enable that freedom and fullness of life for all.[19] The God in whom the church has its being and who dwells in the church is a God utterly active in every dimension of life. And, as revealed in the cross and resurrection, is a God whose life-serving love is indomitable, even when it appears to be defeated.

An ancient faith claim is that God's love in Christ is "flowing and pouring into all things," offering creative, saving and sustaining power for the healing of a broken world.[20] Incarnate mystery lives in and among us as justice making, self-honoring love of neighbor. The church today is called to rekindle that ancient faith claim, to breathe and live in the promise that indeed this God is incarnate in us—mud creatures of the earth[21]–who are gathered to praise God and participate in God's mission. God in us is hungering and hastening toward the restoration of this precious and brutalized world. This vision breathes power to open our hearts and minds to the signs of despair—including our being implicated in ecological and economic violence—without drowning in them. Rather, we can confront those realities on behalf of life abundant for all. In the face of hopelessness or despair, herein lies hope and power for living as the body of Christ.

The cross ends in resurrection hope

The cross communicates hope in the face of despair. For many people, moral inertia in the face of ecological and economic violence is born not of failure to see it, but of hopelessness; the forces of wrong seem

---

[19] Is this "conformation with the form of Christ" a reification of servanthood and self-sacrifice, the state to which women and other marginalized people historically have been thrust and imprisoned? It may appear so, but I think not. Bonhoeffer's theology opposes the assumption that one sector of society is primarily to serve the other. For him, "being for one another" is in the context of also "being with one another." It is not servanthood that is elevated but daring to stand for life in the face of ecological or economic violence—despite the risks entailed. This is the work of people woven by the Spirit into a body in which all give and all receive.

[20] Luther, *LW* 26, as cited by Larry Rasmussen, "Luther and a Gospel of Earth," in *Union Seminary Quarterly Review* 51, no. 1–2 (1997), p. 22.

[21] "Mud creatures," is the English translation of the Greek term used by Irenaeus of Lyons to translate the Hebrew word usually rendered "Adam" in the Genesis creation stories.

too powerful for human beings to impact. Despair is sown by a deep sense that things will continue as they are in this world.

The cross and resurrection promise otherwise. The power of God liberating all of creation from the bonds of oppression, destruction and death is stronger than all that would undermine God's promise of abundant life for all. Soul searing, life shattering destruction and death are not the last word; the end of the story is resurrection. The last word is life raised up out of brutal death. In the midst of suffering and death—whether individual, social, or ecological—the promise given to the earth community is that life in God will reign.

This message of hope also bears danger. It could lead people to abdicate responsibility, leaving it in "God's hands." Bonhoeffer's dialectic between ultimate trust in God, and unwavering critique of liberal Christianity's *deus ex machina*, demands otherwise.[22] His ethic of responsible action to disclose and confront evil is grounded in absolute dependence on God and trust in God. "I believe that God can and will bring good out of evil," he writes, "even out of the greatest evil. For that purpose [God] needs [people] … ."[23] Yet, Bonhoeffer denounces the proclivity to reduce God to "a machine for fixing life's problems" or to expect that only God needs to act. He insists that God's power on behalf of the world is found in God's embodied presence in and with human beings who act responsibly for the sake of life.

Christ present in "all things"

Many streams of Christian tradition have affirmed the *mysterium tremendum* that God dwells within not only human beings, but all creatures and elements. As Martin Luther put it, "… [T]he power of God … must be essentially present in all places even in the tiniest leaf."[24] God is "present in every single creature in its innermost and outermost being … ."[25] God "is in and through all creatures, in all their parts and places, so that the world is full of God and He fills all … ."[26] "… [E]verything is

---

[22] Bonhoeffer, *op.cit.* (note 9), pp. 281–82, 341, 361.

[23] *Ibid.*, p. 361.

[24] Luther, *op. cit.* (note 7), p. 57.

[25] *Ibid.*, p. 58.

[26] WA 23.134.34, as cited by Rasmussen, *op. cit.* (note 20), p. 22, citing Paul Santmire, *The*

full of Christ through and through ... ."²⁷ "[A]ll creatures are ... permeable and present to [Christ]."²⁸ "Christ ... fills all things. ... Christ is around us and in us in all places. ... he is present in all creatures, and I might find him in stone, in fire, in water ... ."²⁹

If indeed Christ fills earth's creatures and elements, then the earth now being "crucified" by human ignorance, greed and arrogance is, in some sense, also the body of Christ. Followers of Jesus the Christ, in every age are charged to ask Bonhoeffer's question, Who is Christ for us today? Where is the cross today? Where are we lured into denying Christ crucified today? If earth, as God's habitation, as body of Christ, is cruciform, and if believers took seriously this christological claim, might we be motivated to treat this earth differently?

Furthermore, if God is boundless justice seeking love, living and loving not only in human beings, but also in the rest of creation, then other-than-human creatures and elements also embody God's intention that all of creation flourish. Earth embodies God, not only as a creative and revelatory presence, but also as a teaching, saving, sustaining and empowering presence—as agency to serve the widespread good. How might moral agency—the power to resist social and ecological destruction and to move toward just, sustainable life ways—be fed and watered in human beings by this God presence and God power coursing through "all created things"?³⁰

These two notions—of Christ crucified in a crucified earth, and of God's saving power dwelling within the created world—may motivate us and empower us for a long, uncharted journey. It is the journey toward a world in which humankind is no longer toxic to our planetary home and in which none amass wealth at the cost of others' impoverishment. Pursuing these theological possibilities at the intersection of cross and indwelling Presence, may be key to a theology of the cross capable of enabling moral agency in the face of ecological and economic violence today.

Travail of Nature: The Ambiguous Ecological Promise of Christian Theology (Philadelphia: Fortress Press, 1985), p. 129.

[27] Santmire, *ibid.*, p. 387.

[28] *Ibid.*, p. 386.

[29] Luther, "The Sacrament of the Body and Blood of Christ—Against the Fanatics," in Timothy F. Lull (ed.), *Martin Luther's Basic Theological Works* (Minneapolis: Fortress, 1989), p. 321.

[30] Luther, "Confession Concerning Christ's Supper," in Lull, *ibid.* (note 29), p. 397.

In sum

Today, ways of life that we assume to be "good" are destroying the earth's capacity to sustain life as we know it, and are generating a massive gap between those who have too much and those who have too little for a life with dignity. In this context, Lutheran and other religious traditions are called to the great challenge of our era: forging ways of life that nurture rather than threaten the earth's health, and that simultaneously counter the structures of oppression that produce excessive wealth for a few at the expense of the lives of many others.

This "great work" is inspired and empowered by a living theology of the cross.[31] We have asked here, How might the cross enable eco-reformation toward lifeways that nurture the earth's health and that build economically just relations with neighbors far and near? In response, we have noted four key factors that may hinder that movement and breed moral inertia: (1) the tendency not to recognize our own participation in social structural sin; (2) the tendency not to recognize who we are called and empowered to be as participants with God in God's work on earth; (3) a sense of powerlessness in the face of systemic forces that seem beyond human agency to impact; and (4) an anthropocentric understanding of God's indwelling presence. Finally, we have identified how a Lutheran theology of the cross can overturn those barriers, turning to Dietrich Bonhoeffer and to the ancient Christian understanding that God abides within God's creation. The latter opens doors to sources of moral motivation and agency that have been obscured by more anthropocentric notions of God's indwelling presence. Bonhoeffer's lived theology of the cross and resurrection reveals who we are as both perpetrators of systemic sin and bearers of God's liberating and healing love. And it offers assurance that—by the grace of God—love will ultimately triumph over the sin.

---

[31] "Great work" is the phrase coined by Thomas Berry to describe the work of creating a sustainable relationship between the human species and planet earth.

Invoking the Spirit amid Dangerous Environmental Change

Sigurd Bergmann

If the Spirit of the Triune is a Spirit who gives life—how can Christians and churches follow the Spirit to wherever it might blow? How might climatic and environmental change today mirror the drama of the history of salvation? Are humans as well as the whole suffering, groaning creation "hoping for liberation"? Where does the Spirit of life-giving remembrance and liberation "take place" today?

The increasing consciousness about ongoing environmental change, locally as well as globally, is challenging much in Christian theology. Human-caused environmental change, as science clearly points out, is becoming dangerous, that is, irreversibly changing climatic and other conditions for life systems. For vulnerable human ecologies, such dangerous change substantially threatens survival, while it challenges wealthy nations to change their infrastructures and lifestyles. Furthermore, dangerous environmental change increases economic, social and military tensions in many regions, and is already victimizing a large number of people in economically weak and vulnerable areas of the world. The extinction of certain species is expected to increase rapidly, ecological diversity is reduced and safe water will become increasingly scarce.

The fact that environmental change is anthropogenic—a human-caused social and historical construction of an expanding capitalist, technocratic and unjust world system—is what sharpens the religious dimension even more. How can God be the Creator, Sustainer and Liberator, and how can human beings be understood as being in the image of God, if they destroy the gift of life? Where is God in this? Do climate change and the environmental catastrophe indicate a punishment for human sin? Is God angry? How should we imagine the Divine? Is God absent or present with the suffering?

Environmental change not only transforms the conditions of life, but also radically changes culture, religion and the conditions for faith. Therefore, the question for faith communities of different confessions is how religion itself can change in light of this. What might believers contribute to creative adaptation to environmental change?[1] Here I will focus on spatiality, climate politics, eco-justice and the remembrance of suffering.

A spatial turn

In the face of dangerous changes in climate and water systems, I propose that Christianity needs to accelerate its turning toward "space." In relation to the challenges of that time, twentieth-century theologians reflected mainly in terms of time and history. But climatic changes clearly turn our focus to the spatiality of creation. As one common gift of life, creation reveals its glory in the complexity, diversity and interconnectedness of life systems, in one single planetary space for all.[2] "Earth is our home," states the declaration of the Earth Charter.[3] This simple statement summarizes a deep wisdom that has been guarded by religions for many ages. "The earth is the Lord's" (Ex 9:29; Ps 24:1). Paul makes very clear that the earth and humans as God's icons are interconnected in a communion of suffering and hope for liberation (cf. Rom 8:21).

Theology's "spatial turn"[4] was initiated in the last few years and will undoubtedly accelerate due to dramatic changes in our common planetary space. Thus, the challenge to faith traditions is to explore and interpret how the life-giving and all-embracing space given by the Creator contrasts with the global space where risks and damages are being experienced in violent and unjust ways. Is God's good, all-embracing space turning into a catastrophic space where some are victimized for the sake of the survival of others? How does God's love for the poor relate to situations where the most vulnerable become the

---

[1] For a deeper reflection see also the contributions to a trans-disciplinary workshop in October 2008, arranged by the EFSRE and PIK program, at **www.hf.ntnu.no/relnateur/index.php?lenke=ridecc.php**, forthcoming in Sigurd Bergmann and Dieter Gerten (eds), *Religion in Dangerous Environmental Change*, Studies in Religion and the Environment 2 (Muenster/Berlin/Zurich/Vienna/London: LIT, 2009).

[2] On the theology of "gift events" in our context, see Anne Primavesi, *Gaia and Climate Change: A Theology of Gift Events* (London/New York: Routledge, 2009).

[3] At **www.earthcharter.org/**, accessed June 2009.

[4] Sigurd Bergmann, "Theology in its Spatial Turn: Space, Place and Built Environments Challenging and Changing the Images of God," in *Religion Compass 1* (3/2007), pp. 353–79.

most victimized? Why does climate justice need to become a central Christian virtue today?[5]

Scenes of catastrophic environments are not strange in the history of Christianity. They are depicted, for example, in the illustrated 1534 Luther Bible, where the destruction through the flood is dramatically contrasted to the Creator's love for creation. The purpose in these and most other stories of God's angry reaction to human injustice and sin is pedagogical. According to the second-century church father Origen, creation should be understood as a divine school, a *paidagogia*, a place for humans who are not yet mature and need to be developed or educated through experience and reflection.[6] Nature's responses to human beings increasing emissions, extinguishing species and mismanaging the water can be regarded as a message from the Creator to humans. "Creation [itself] preaches," as Gregory of Nazianz, the fourth-century church father and poet, put it.

This does not mean that science is turned into a religion, as some skeptics have accused climate scientists, but the results of climate science can indeed be interpreted within a spiritual framework. Life systems have the capacity to relate and react in one common Gaia earth system, as a part of God's good creation. From such a perspective, dangerous environmental change offers insights into the limits for managing the earth, and locates the roots of the crisis in spiritual as well as social and physical failures. That which threatens life also threatens creation, and at its outmost, authentic belief in the Creator. Words and worlds are intimately interconnected; images of the Divine are interconnected with the physical surroundings and cultural environments. The image of God and the image of nature are indissolubly interwoven.

The early twentieth-century dispute in Europe between Futurist and Dadaist artists provides a microcosm of the still ongoing controversy about modernity, as reflected today in discourses about the climate. While Futurists were glorifying speed, progress and technical innovation and were enchanted by its potential to change urban lifeworlds, the

---

[5] Cf. Michael S. Northcott, *A Moral Climate: The Ethics of Global Warming* (London: Darton, Longman and Todd, 2007), pp. 7, 118; David Atkinson, *Renewing the Face of the Earth: A Theological and Pastoral Response to Climate Change* (Norwich: Canterbury Press, 2008), pp. 93ff.

[6] Cf. Franz Xaver Portmann, *Die göttliche Paidagogia bei Gregor von Nazianz* (St. Ottilien: Eos Verlag, 1954); Sigurd Bergmann, "Atmospheres of Synergy: Towards an Eco-Theological Aesth/Ethics of Space," in *Ecotheology* 11 (3/2006), pp. 326–56, here p. 338. On God's skill to transform the bodily created and to promote spiritual education through human bodily being, see Origen, *De Principiis*, III, 6–9.

Dadaists objected to the fascist glorification of strength and progress and the technocratic attitude toward nature. While the one depicted a world of change and construction, the other predicted a world of waste and destruction.

One of the driving forces behind civilization was the development of urban space, as it began more than 10,000 years ago.[7] The accelerating process of urbanization is now turning the whole planet into one single "post-metropolis." Today, the majority of the world's population live in urban areas. This affects the reshaping of landscapes and regions worldwide. With regard to climate change, city space promotes and accelerates fuel-driven, profit-oriented industrialization and motorized mobility, which produce large amounts of emissions. In countries of the North as well as in the South, "hypermobility" has reached a stage that threatens the whole system of social and environmental planning and even the very roots of democracy. The French philosopher, Paul Virilio, characterizes the present state of culture as a standstill caused by increasing acceleration.[8]

In this context, religious attitudes can take two directions. Either they will follow the Futurist glorification and politics of speed and progress, or they will cultivate the Dadaist desire for alternative ways to organize urban life. A central question is how a city can be turned into a habitable place and what religion can contribute. What are the criteria for a place being habitable, and how can this space be designed in light of what we know about emissions and energy saving?[9]

How we perceive our human constructed or "built" environment is at the core. Spiritual values and ecological justice need to be significant factors in urban transformation. Such an approach, which I call "aesth/

---

[7] Edward W. Soja, *Postmetropolis: Critical Studies of Cities and Regions* (Oxford: Blackwell, 2000), p. 35.

[8] Paul Virilio, *Fluchtgeschwindigkeit: Essay* (Frankfurt/M.: Fischer, 1999); Cf. Hartmut Rosa, *Beschleunigung: Die Veränderung der Zeitstrukturen in der Moderne* (Frankfurt/M.: Suhrkamp, 2005); Sigurd Bergmann, "The Beauty of Speed or the Discovery of Slowness–Why Do We Need to Rethink Mobility?, in Sigurd Bergmann and Tore Sager (eds), *The Ethics of Mobilities: Rethinking Place, Exclusion, Freedom and Environment* (Aldershot: Ashgate, 2008), pp. 13–24.

[9] Cf. Tim J. Gorringe, *A Theology of the Built Environment: Justice, Empowerment, Redemption* (Cambridge: Cambridge University Press, 2002); Pauline von Bonsdorff, "Habitability as a Deep Aesthetic Value," in Sigurd Bergmann (ed.), *Architecture, Aesth/Ethics and Religion* (Frankfurt/M., London: IKO-Verlag für interkulturelle Kommunikation, 2005), pp. 114–30. On architecture and urban planning, cf. also Sigurd Bergmann (ed.), *Theology of Built Environments: Exploring Religion, Architecture and Design* (Piscataway NJ: Transaction Publishers, 2009).

ethics," implies the integration of ethics in our bodily ways of being. Moral ways of acting need to be integrated into our bodily modes of perceiving each other and living in our surroundings. As climate changes, our knowledge and interpretation of climate need to be in harmony with our bodily perceptions of what climate does to us.

Following Jakob von Uexkuell's "circle of function" *(Funktionskreis)*, the single organism's lifeworld not only is connected in a two directional way between the *Merkwelt* (way of viewing the world) and *Wirkwelt* (universe of action), but climate change appears as a huge challenge to visualize the flows of this circle between our perception and our global and local action. Global space, with all its local places, needs to be revisioned as a cosmos of synergies, where the symphony of what surrounds us moves back into our consciousness. The complexity of climate studies and the phenomena of interconnectedness can catalyze such a changing worldview of lived space. Religion as a force of productive imagination *(produktive Einbildungskraft)* is crucial in this reimaging of our environment.

Edward Soja has made a theoretically useful distinction between three types of space: physical, imagined and lived space.[10] The concept of "lived religion" corresponds to this concept and offers a means for studying climatically caused religious change in today's urban spaces.[11] Roy R. Rappaport has described religion as a "fabrication of meaning." He differentiates between a scientific and a religious mode of approaching reality, and argues for a synthesis between them.[12] For him, the basic tension in human life lies between making meaning with regard to our environment and explaining the laws of this environment.[13] Applying this to climate change, there is a tension between rationally monitoring and explaining our changing environment on the one hand, and making meaning in relation to it on the other hand. Both are needed.

It is my hypothesis that a theological construction of meaning in the climate change discourse is an important opportunity for such a

---

[10] Edward W. Soja, *Thirdspace: Journeys to Los Angeles and Other Real-and-Imagined Places* (Malden/Oxford/Carlton: Blackwell, 1996, reprinted 2004).

[11] Cf. Sigurd Bergmann, "Lived Religion in Lived Space," in Heinz Streib, Astrid Dinter and Kerstin Söderblom (eds), *Lived Religion: Conceptual, Empirical and Theological Approaches, Essays in Honor of Hans-Günter Heimbrock* (Leiden/Boston: Brill, 2008), pp. 197–209.

[12] Roy R. Rappaport, *Ritual and Religion in the Making of Humanity* (Cambridge: Cambridge University Press, 1999).

[13] *Ibid.*, chapter 14.

synthesis. This does not mean that everything in science that fuels a religious attitude to life is good, or that every scientific analysis should be transformed into a policy generating strategy. On the contrary, human beings are the agents of perception, knowledge, wisdom and meaning making; it is they who change their lifestyles and patterns of production and consumption. In this context, theology must guarantee that human beings are the agents of social transformation, and not to be treated as objects. If the earth, our home, is to remain a habitable place where all can live, spiritual perceptions, perspectives and practices are crucial for developing means of creative adaptation to our changing environment.

Climate regimes and faith communities

Science and politics are the dominant power constellations—or "climate regimes"—that presently control the public discourse on climate change. Even in the European Union, where "citizens' governance" claims to be a priority, quasi democratic processes of decision-making "sub-political alliances" (Ulrich Beck)[14] do not follow the principle that power comes from the people. On the European as on the global scale, it is unclear how the church as a *communio sanctorum* should participate in political negotiations.

Climate change sharpens the challenge to speak with one religious voice and to broaden the discourse. The voices of the poor and other victims of climate change especially need to be represented. One hopeful sign for the emergence of a common voice of different religions is the manifest from the 2008 Interfaith Climate Summit in Uppsala, even though this is still too general a document.[15]

Should only scientists develop the analyses and decide on strategies of adaptation and mitigation, which economists and politicians then apply and impose on the populace? What would a good "practical discourse" (Jürgen Habermas)[16] about dangerous environmental change be like?

---

[14] Beck is professor of sociology at the Ludwig-Maximillian University, Munich, and visiting professor at the London School of Economics.

[15] "The Uppsala Interfaith Climate Summit Manifesto, "Hope for the Future!," initiated by the archbishop of the Church of Sweden, can be downloaded from, **www.svenskakyrkan.se/default.aspx?di=173305&ptid=0**, accessed June 2009.

[16] Habermas is an influential German sociologist and philosopher.

At present, faith communities and social movements are all too often reduced to being tools for mobilizing people.[17] But why would we expect solutions to these deep problems to come from those same systems that over the past 150 years have produced the developments that have accelerated global warming, deforestation, the extinction of certain species and the economization and "technification"[18] of lifestyles?

I do not want to minimize the importance of ongoing activities of scientific, political and business leaders. However, a task of Christian theologians is to "convert" the powerful to confess their guilt or complicity and to repent. Without such public confession and expiation, which are credible to the victims, no real forgiveness can take place. Are the churches prepared to call for such a public process, as occurred earlier in church history as well as more recently in South Africa?[19]

Because the relationship between climate and religion is much deeper than usually assumed, contributions and interventions from faith communities are crucial for navigating into our common future. Those who hold power, such as transnational corporations, are much more aware of the importance of symbolic values and performances in the public sphere than are most church leaders. This is not surprising at a time when money has become such a religious artifact, working only by faith in its exchange value. Financial markets become very nervous whenever religious arguments are made that it is "money [which] makes the world go around." Money also makes the climate go around, which turns us to the painful question of how we can believe that both faith in God and faith in money can be combined. "No one can serve two masters; ... You cannot serve God and wealth" (Mt 6:24). How can this inform a pastoral theology for churches in times of climate change?

Climate change clearly reveals the historical change of relations between humans and their environments. The last 10,000 years of human survival and civilization have been possible only because of the stabilization of climatic conditions. We wonder whether we have entered the end of this period. The conclusions of the IPCC reports are

---

[17] At **http://climatecongress.ku.dk/**, accessed June 2009.

[18] Cf. Gernot Böhme, *Invasive Technisierung: Technikphilosophie und Technikkritik* (Zug: Die Graue Edition, 2008).

[19] Cf. Ernst M. Conradie, "Schuld eingestehen im Kontext des Klimawandels," in *Salzburger Theologische Zeitschrift 12* (1/2008), pp. 48–74, and Ernst M. Conradie, "Healing in a Soteriological Perspective," in *Religion & Theology: A Journal of Contemporary Religious Discourse 13* (2006), pp. 3–22.

obvious enough: without mitigation and adaptation to new conditions, civilization, as we know it today, cannot survive. Neither can globalized business or the present economic world system. This challenges faith communities, as well as other bodies, to resist the temptation of reducing climate change merely to stabilize global capitalism. This is the driving force behind reports that simply deal with climate change in terms of financial implications.

In spite of enormous scientific and technical innovations, the managers of modern society have not only lost control but also the ability to see what climate is doing to us at present. They strive to replace uncertainty with new certainties through old empiricist methods. Their vision is of a better, even more sublime "management of the planet," a vision that seems to be dangerous and, in the long run, works against developing an alternative agency of nature.

Such an alternative would need to be grounded in the spiritual awareness of life as a gift, of nature as a complex ecology with intrinsic value, and of an all-embracing space that should be thought of in terms of metaphors such as home, garden and body, rather than as a machine, system or market. Religious traditions have a wide range of metaphors for life, nature and space to contribute to how we perceive and conceive, including in science. Such metaphors and images influence our inner images, perceptions and concepts of reality in fundamental ways. They are invisible driving forces behind every theory that has emerged in science.[20] The notion of evolution, for example, was unconsciously transferred by Darwin into his theory from the liturgical unrolling (*evolvere*) of the holy books in the synagogue. "In the beginning is the icon" and, as John made clear in the beginning of his Gospel, also the *logos*.[21]

Another contribution from religion to climate research is to question the attempts to reduce and eliminate all uncertainty. Not every certainty is good, and not every uncertainty is dangerous. What kind of uncertainty leads to deep wisdom about life, and what kind of uncertainty should we try to reduce? Could the normal scientific intention to reduce any uncertainty in weather prognosis and climate models lead to a dangerous attitude and the practice of simple managing? Or, could it assist in developing sensitive attitudes of awareness and learning

---

[20] Hans Blumenberg, *Die Lesbarkeit der Welt* (Frankfurt am Main: Suhrkamp, 3rd edition, 1993), p. 19.

[21] Cf. Sigurd Bergmann, *In the Beginning is the Icon: A Liberative Theology of Images, Arts and Culture* (London: Equinox, 2009), chapter I.

about the unpredictable dimensions of nature, in ways similar to those of indigenous hunters?

Religions specialize in dealing with uncertainty, and spiritually differentiated uncertainty is also needed in empirical science. Scientists can learn from the historical failures of theologians in antiquity and medieval times. If, on the one hand, theologians would try to make certain prognoses about God's being and acting in the world, they violate the mystery of the Triune in ways that may destroy faith. On the other hand, if theologians avoid interpreting God in the modern world, they only mystify the places and situations where the Spirit acts, liberates and gives new life. The balance of certainty and uncertainty is a necessary aspect of spirituality in general. Already in the fourth century, theologians such as Gregory of Nyssa and Gregory of Nazianz framed the so-called limits of rational certainty in terms of "apophaticism."[22] We should be silent about God's essence, which we cannot know, but speak positively about God's actions in creation, which we can experience bodily. The same is also true for nature. Nature in general should not be forced into one single frame of scientific theory but its many modes of being should be reflected in concepts that convey sensibility, awareness and spiritual respect.

In this regard, environmental change leads human beings to reflect anew on their own identity and relationship to nature. How are we located in what the twentieth-century German language poet, Rainer Maria Rilke, called *Weltinnenraum* (inner-world space)? While the early phase of climate research tried to postulate about certain developments, nowadays it seems much more important to point to the direction where we need to accept, understand and value the intrinsic uncertainty implicit in earth systems. The distance between science, religion and culture thus shrinks.

The wisdom of ecological justice

The problems emerging from climate change are located in a broader history of human agency with nature. Modernity and its many "blessings" have led to an historically unique human disembeddedness from

---

[22] Cf. Sigurd Bergmann, *Creation Set Free: The Spirit as Liberator of Nature*. Sacra Doctrina: Christian Theology for a Postmodern Age 4 (Grand Rapids: Eerdmans, 2005), pp. 339ff.

natural environments. Scientific and technological developments, combined with a deregulated global financial market, have thrown us into a state of affairs that requires a kind of cultural revolution. The roots of this global ecological crisis, debated now for over thirty years but observed already in the early nineteenth century, can be described as sociocultural as well as spiritual.

Environmental problems penetrate many sectors of society, in many different patterns of production and consumption, or what the German scholar Ernst Ulrich von Weizsäcker rightly declared to be the millennium of the environment after the last millennium of the economy.[23] Political scientists point to the emergence of a kind of global "ecological citizenship."[24] The process of a transition to an environment based global culture is only beginning. A widely accepted consensus is that socially produced problems in the interaction of human beings and nature cannot be solved only through technical or social engineering. Environmental problems are in the first instance problems of human and cultural constructions; technical and material forms of a non-sustainable eco-management are interwoven with and express human values, attitudes, ultimate concerns and power realities.

Established religious traditions and institutions in Europe have critiqued the dependency on fossil fuels for mobility.[25] The criticism of consumerism as an idolatry has been highlighted in combination with a long-term commitment to the oceanic lifeworld. The Ecumenical Patriarch of Constantinople Bartholomew I and other church leaders have published well-informed public declarations. For some time, the United Nations Environment Program (UNEP) has included religious leaders in their work.

There is a strong basic desire for an alternative way of living, which produces a creative spiritual force for social transformation. In medieval Christianity, monks talked about departing from this world: *exire de saeculo*. Put in early Christian language, climate change makes it

---

[23] Ernst Ulrich von Weizsäcker, *Earth Politics* (London and New Jersey: Zed Books, 1992), pp. 3ff.

[24] Cf. Deane Curtin, "Ecological Citizenship," in Engin F. Isin and Bryan S. Turner (eds), *Handbook of Citizenship Studies* (London: Sage, 2002), pp. 293–304.

[25] World Council of Churches, *Mobilität: Perspektiven zukunftsfähiger Mobilität* (Geneva: World Council of Churches, 1998) and World Council of Churches, *Mobile–but not Driven: Towards Equitable and Sustainable Mobility and Transport* (Geneva: World Council of Churches, 2002). Cf. also Jutta Steigerwald, "Walk the Talk: Mobility, Climate Justice and the Churches," in Bergmann and Sager, *op. cit.* (note 8).

necessary to mine our vision of the Spirit who acts for "the life of the coming world," as formulated in the Nicene Creed. This also calls for love of the poor as the highest kind of love of God. As Michael Northcott has shown, climatic change deepens the injustice between rich and poor, and without a strong participation of local communities climate politics is doomed to fail.[26]

Since the times of the Egyptian goddess Ma'at ("justice"), in the world of religions, the quest for justice has carried a sense of the divine on earth. Eco-management has been contrasted with eco-justice.[27] Unjustly distributed risks test the vision of a just world. The participation of faith communities is needed in adapting to climate change, if social life is to be reconstructed with ecological justice for all.

In the context of climate change, the challenge is to mobilize spiritual energies and to connect them with scientific information and a social arena where experimenting with alternatives can take place. In 1964, the deep ecologist, Arne Næss, coined the term "ecosophy" for this, where ecology as a scientific activity and sophiology as a kind of rational belief system interact without being dissolved into each other.[28] Nicholas Maxwell has expressed a similar approach in his criticisms of the philosophy of knowledge, when it does not take into account the self-interests of living beings. Instead, he argues, knowledge should be replaced with wisdom. This is truth seeking that negotiates why we should do or abstain from doing something.[29]

When approaching climate change from a theological perspective, the challenge is to explore the desire and spiritual driving forces that are implicit and explicit in environmental commitments, and to widen the present discourse with a wisdom framework for reflecting on why we should act or abstain. A worldview based on economic principles only allows the argument that the cost will be greater if we do not respond today. A wisdom based discourse, however, explores reasons for why local and translocal perceptions need to be included, along with

---

[26] Northcott, *op.cit.* (note 5).

[27] Cf. Dieter Hessel (ed.), *After Nature's Revolt: Eco-Justice and Theology* (Minneapolis: Fortress Press, 1992).

[28] Arne Næss, *Ecology, Community, and Lifestyle: Outline of an Ecosophy* (Cambridge: Cambridge University Press 1989, pp. 37–40. (Shortened and revised edition of *Økologi, samfunn og livsstil:Utkast til en økosofi*, Oslo: Universitetsforlaget, 1974).

[29] Nicholas Maxwell, *From Knowledge to Wisdom: A Revolution in the Aims and Methods of Science* (Oxford: Blackwell, 1984).

practical knowledge and skills for surviving with dignity in sustainable environments.

Reflecting on values, lifestyles, cultural practices, worldviews, and life interpretations becomes necessary in seeking solutions. Religions have universal and translocal validity. The visible hides that which is invisible and powerful, such as in animism, totemism or indigenous religion. Asian religious traditions have complex structures of divinities or energies that shape and rule over cosmic processes. In contrast, monotheist religions in the West regard the one God as the Creator, Sustainer and Liberator of the world. Yet, creation is filled with signs and traces of God. In this sense, nature is revelatory. The earth and its inhabitants are gifts that humans can never manage on their own. "The earth is the Lord's" (Ex 9:29; Ps 24:1). Similar expressions are found in many religions, such as in the ecological awareness of Buddhism.[30] Muslims have established the dynamic Islamic Foundation for Ecology and Environmental Sciences (IFEES),[31] and have offered eco-theological interpretations of the Koran and other traditional sources.

A traditional response to threatening changes in the natural environment is that God (or the spirits) are reacting to sinful human practices. For example, Gregory of Nazianz interpreted droughts causing starvation as a reaction of the trinitarian God who "preaches through creation."[32] Augustine imaged nature as a *liber naturae* in addition to the Bible (*liber scriptura*). For him and the entire medieval tradition until the Renaissance, nature and the Bible represented two different but equal books in which one could decipher the law and the love of the Creator toward his creatures. For the medieval physician Paracelsus (1493–1541), natural science was an hermeneutics of the Creator's signs in creation. Medieval science developed as an interpretation of this invisible, uncreated God in nature and its laws. While antiquity used rational explanation as a tool for wisdom, during the Middle Ages, theoretical and practical strategies began to be developed to control the environment, a process that was furthered during the Enlightenment. While "enlightened vitalism"[33]

---

[30] Venerable Jinwol, *Seon Experience for Ecological Awakening*, forthcoming, in Sigurd Bergmann and Kim, Yong-Bock (eds), *Religion, Ecology and Gender: East-West Perspectives*, Studies in Religion and the Environment 1 (Muenster/Berlin/Zurich/Vienna/London: LIT, 2009).

[31] At **www.ifees.org.uk/**, accessed June 2009.

[32] Gregory of Nazianz, Or. 38.13, 16.5, 6.14. Cf. Bergmann, *op. cit.* (note 22), p. 109.

[33] Peter Hanns Reill, *Vitalizing Nature in the Enlightenment* (Berkeley/Los Angeles/London: University of California Press, 2005).

sought to understand vital forces in a sensitive way, modern epistemology has set creation free from all kinds of ideological limitations other than the ultimate concern for accumulating profit. "It is hardly too much to say that scientific and technological research have made possible all our current problems."[34]

Consequently, modern science and technology can be interpreted as instruments for salvation in a religious key, even if they claim to have secular neutrality. Climate research, as located in environmental science, breaks radically with such a code of science due to an alternative understanding of what is normative. While science seeks insights for the sake of human interests, climate and environmental science seeks insights for the best of the earth itself, including its inhabitants. The latest approach to climate geopolitics expresses this in the equal right to emissions necessary for every global citizen to survive.[35]

However, what is the best for all and how it can be achieved is a question that should include the participation of all affected. Global institutions need to offer space for such dialogue. Otherwise, consensus about the good life for all cannot be achieved. Might a consensus among religious institutions and communities about climate change as a call to spiritual change help further this? Could it contribute to deeper formation of meaning than the economic management of climate and nature? Does the economic "uncertainty" of environmental processes present a threat or a challenge?

Remembering the above historical connections could contribute to a new ethics of respect. Religions mirror this deep connection of climate environments and faith. The story of Genesis presupposes the existence of a garden in the land of the two streams, which was among the first regions stabilized by the new climate conditions. The story of the great flood reflects the memories, which at the time were still present after experiences of unpredictable climate changes. The prophet Jeremiah offers one of the oldest narratives of a dramatic crisis of culture and nature, which at the time was also mirrored in the drama of climate (cf. Jer 18:14–17).

---

[34] Nicholas Maxwell, *Are Philosophers Responsible for Global Warming?* Email letter to the mailing list, **PHILOS-L@Liverpool.ac.uk**, 13 October 2007.

[35] Equal rights to emissions for every global citizen have been proposed by the German Chancellor, Angela Merkel, in September 2007. Such a proposal would, however, need to acknowledge the difference between luxury and ordinary life emissions. The principle of equal rights is also applied in Inge Johansen's report for the Norwegian Academy of Technological Sciences (NTVA), *Ethics of Climate Change: Exploring the Principle of Equal Emission Rights* (Trondheim, 2007).

In the world of religions, dramatic environmental changes have usually been interpreted as signs from the Creator, who does not plague creatures arbitrarily, but sets signs into the sky in order to make them aware of their failures, sins and crimes against God's laws in nature. Dramatic environmental change therefore has been interpreted as a means for people to find their way back to an ecologically right path. This foundational code has been present in many religions and cultures throughout time. Hence, environmental ethics are not a modern invention but live from roots of older religious traditions.

Remembering the suffering and liberation of creation

Ethical decisions are not simply made by theoretical moral arguments according to universal principles. An important source of strength in the context of dangerous environmental change can be found in the process of remembrance, especially of the suffering. As evident already in this book, climate change is not just a coming threat but has already caused a large number of victims. Again, the poor carry the greatest sufferings. The memories of suffering are embedded in local perceptions and practices.

The theologian Johann Baptist Metz, reflects on the significance of the remembrance of the suffering.[36] Remembering the suffering is necessary in order to find and apply processes of actual liberation from powers of teleological and technological thinking. The need to integrate the remembrance of the suffering is crucial for the process of urban planning,[37] and also in relation to climatic change. Remembrance of the suffering caused by climate change can contribute to environmental planning and to a creation of meaning in lived space and lived religion.

For Christians, the foundation of such an hermeneutics of suffering is our remembrance of Christ's sufferings. The Holy Spirit makes this remembrance possible. In this, the tradition of the oppressed is key. This tradition is found "only among those who are poor and despairing right

---

[36] Johann Baptist Metz, "Erinnerung des Leidens als Kritik eines teleologisch-technologischen Zukunftsbegriffs," in *Evangelische Theologie* 32 (1972), pp. 338–52.

[37] Sigurd Bergmann, "Making Oneself at Home in Environments of Urban Amnesia: Religion and Theology in City Space," in *International Journal of Public Theology* 2 (2008), pp. 70–97.

now and only at the places where they suffer–only near to them."[38] It is precisely their situation that provides them with a powerful impetus to believe, to hope, to experience and to reflect on God's liberation. The challenge is to respond to dangerous environmental changes in ways that respect or benefit the memory of those who have been thus excluded.

This remembrance of the suffering also affects the understanding of Christian eschatology. It encourages us to understand more deeply the Spirit's work in space and time. Eschatology often mistakenly separates the past from the future. The Holy Spirit, however, creates not only the life of the world to come, but also nurtures the present through remembrance. In this sense, the Spirit is the agent of tradition and the subject of remembering and handing over. The memory of local Christian experiences, artifacts and practices represents a dynamic synthesis of continuity and change. The Holy Spirit keeps theological interpretations of life open for the future as well as keeping the process of remembrance, the writing of history and historical imagination open for the past.

While such an insight has been central in Asian and African religions, in which ancestors have essential significance for the present, the dignity of the power of the past has been expelled by the modern West's powerful "myth of progress" and its shift to regard history as a human product. Because the Spirit works to appropriate both the past and the future, both need to be treated not as predictable or reducible but as spiritually open to God's work. The past as well as the future continuously offer thresholds for the Creator to pass over into creation. Pneumatology, therefore, interprets how the advent experience of waiting for God to come, and the historical experience of remembering the God who has come, both come together to create an open attitude toward life as a gift flowing from the past as well as into the future.

To do this, the Spirit needs to be understood as a movement. The Spirit, who goes between, moves through the borders of space and time. Such a perspective leads faith communities to ask, Where does the Spirit's life of the coming world emerge? How does the Holy Spirit "take place" in climate change? How do communities of believers interact with a liberating God in lived space? How do they transform built environments into habitable space, flourishing with justice both for humans and for others?

---

[38] Ottmar John, "Die Tradition der Unterdrückten als Leitthema einer theologischen Hermeneutik," in *Concilium* 24 (6/1988), pp. 519–26, here p. 524.

The reason why religion is necessary to complement science, politics and technology is simple. In dialogue with Carl Friedrich von Weizsäcker, the philosopher Georg Picht asked, How can science be true if its applications destroy its own object, nature? Until today, the question remains unanswered. One could continue to ask, How could technology be true if its application in a world of artifacts, the "second nature" (Walter Benjamin),[39] leads the first nature into collapse? And how, finally, can economic theory be true if its applications, mainly with neoliberal models, lead to an increasing impoverishment of nature and world populations?

The answers can only be worked out in alliances of scientists and believers who are committed to an exile out of "this world" and empire. Fortunately, small processes of *exire de saeculo* in this sense are already taking place today in politics, the arts, even the World Bank and science and other spheres today. Therefore, the urgency of building alliances is increasing, and faith communities, because of their translocal and transcultural reach, have a unique capacity to further this process.[40]

Could we revise ecclesiology in this sense and reimagine the church as an agglomeration of local places and a global space for creative experiments in the arts of survival in environmentally changing contexts? Where will the life-giving Spirit "take place" in a changing environment?

---

[39] Literary critic and philosopher (1892–1940).

[40] On the challenge of climate change to a revised self-understanding of the churches see Ernst Conradie, *The Church and Climate Change* (Pietermaritzburg: Cluster Publications, 2008).

Appendix

LWF Resolution on Climate Change

At its meeting in June 2008 in Arusha, Tanzania, the LWF Council voted:

"To call upon member churches to engage in and deepen their theological and ethical reflection on the human contribution to climate change (drawing on the resources being developed through the DTS program "Responding Theologically to Climate Change"), recognizing human beings as 'co-creatures' with moral agency rather than claiming the prerogatives of creators, and seeking to learn from Indigenous practices and traditional wisdom for living sustainably as part of Creation.

"To impress upon member churches the critical urgency and unprecedented magnitude of the challenge of climate change and the threat that it poses to humanity and all living beings on Earth.

"To urge member churches to move beyond lamentation to urgent and effective action, especially in relation to reducing the emissions generated by their institutional activities and operations (including travel) and promoting more sustainable lifestyles and behaviors among their members.

"To request the General Secretary and LWF member churches to undertake targeted advocacy actions (together with the WCC and other ecumenical and civil society partners) in appropriate forums – including forthcoming sessions of the UN Climate Change Conference – in order to promote strong political commitments to achieving a 40% reduction compared to 1990 levels of CO_2 emissions by 2020.

"To encourage advocacy by the General Secretary and LWF member churches for rapid transition from coal and other fossil fuels for power generation to non-nuclear renewable energy sources, and for the application of a "carbon tax".

"To call upon the General Secretary, LWF member churches and LWF field programs to consolidate and enhance their efforts to address the impacts of climate change on development and poverty in the most vulnerable communities, and to promote effective national and international responses to climate change adaptation and mitigation.

"To ask the General Secretary to implement a CO_2 emissions compensation system for LWF air travel, using appropriate LWF projects, at the latest by the beginning of 2009, and invite all member churches to use this CO_2 emission compensation system for their air travel.

"To request the General Secretary to take other actions necessary to reduce the carbon footprint of the LWF secretariat, including considering the following actions with regard to Assembly preparation:

- reducing the number of printed documents and offering the option of digital or paper documents;

- providing all digital documents on a memory stick; and

- preparing Assembly delegates and participants for a more digitalized Assembly by organizing as part of the pre-Assembly meetings, training workshops on on-line communication.

"To request the General Secretary to produce and disseminate to member churches a briefing paper on climate change and possible responses by the churches."